Dixie Highway

Dixie

Road Building and the

TAMMY INGRAM

Highway

Making of the Modern South, 1900–1930

The University of North Carolina Press ■ CHAPEL HILL

This book was sponsored by the postdoctoral fellows program at the Center for the Study of the American South, University of North Carolina at Chapel Hill.

Designed and set in Utopia and Gotham types by Rebecca Evans
Manufactured in the United States of America

The paper in this book meets the guidelines for permanence and durability of the Committee on Production Guidelines for Book Longevity of the Council on Library Resources. The University of North Carolina Press has been a member of the Green Press Initiative since 2003.

Library of Congress Cataloging-in-Publication Data
Ingram, Tammy.
Dixie Highway : road building and the making of the modern South, 1900–1930 / Tammy Ingram.
    pages cm
"Sponsored by the postdoctoral fellows program at the Center for the Study of the American South, University of North Carolina at Chapel Hill"—Title page verso.
Includes bibliographical references and index.
ISBN 978-1-4696-1298-0 (cloth : alkaline paper)
1. Dixie Highway—History. 2. Roads—Southern States—Design and construc-tion—History—20th century. 3. Roads—Southern States—History—20th century.
4. Transportation—Social aspects—Southern States—History—20th century.
5. Transportation—Political aspects—Southern States—History—20th century.
6. Express highways—Political aspects—Southern States—History—20th century.
7. Express highways—Political aspects—United States—History—20th century.
8. Southern States—Politics and government—20th century. 9. United States—Politics and government—1901-1953. I. University of North Carolina at Chapel Hill. Center for the Study of the American South. II. Title.
TE25.5.D49164 2014   625.70975´09041—dc23
2013029961

18 17 16 15 14   5 4 3 2 1

For my mother

■ ■ ■

and in loving memory of my father

Contents

Figures and Maps

Figures

Maps

Acknowledgments

I learned how to drive on the back roads of south Georgia, roads that in the 1980s were not entirely unlike the rutted dirt roads farmers had navigated by horse and wagon a century earlier. Even before I could see over the steering wheel of my dad's old one-ton flatbed Ford, I explored the mostly unmarked network of narrow dirt, gravel, and paved county roads around our farm. When farmers passed me in their trucks and tractors, they waved. Once, I encountered the sheriff at a four-way-stop, but he just laughed and wagged his finger. Farmers' kids had special privileges in Seminole County, a sparsely populated peanut- and cotton-farming community where long country roads were the lifelines connecting farm families like mine to markets, schools, hospitals, and each other. While this may not explain entirely my decision years later to write about road building in the early twentieth century, I am certain that it helped me to appreciate how important roads were to the people I write about in this book.

My intellectual journey into road building began in graduate school at Yale, when the North Caroliniana Society at UNC–Chapel Hill granted me an Archie K. Davis Fellowship to begin my dissertation research. Thanks to Harry McKown there for sharing his inexhaustible knowledge of the Good Roads Movement with me and for persuading me that this was a topic worth pursuing. For the next several years, grants and fellowships from the Beinecke Rare Book and Manuscript Library at Yale and the Yale Graduate School supported my research and gave me time to think and write. I am grateful to George Miles and the entire archival staff at the Beinecke for their help, and also to the archivists and librarians at Yale's Sterling Memorial Library. Archivists at the National Archives in College

Park, Maryland; UNC's Wilson Library; the Chattanooga-Hamilton County Bicentennial Library; the Washington Memorial Library and Middle Georgia Archives; and the University of Georgia Libraries were particularly helpful in the early stages of my research. Follow-up trips to the Tennessee state archives in Nashville; the Russell Library at UGA; the Rome-Floyd Records Center in Rome, Georgia; and the Georgia Department of Archives and History in Atlanta allowed me to complete revisions.

I am indebted to Dawn Hugh and the staff at HistoryMiami (formerly the Historical Museum of South Florida) for assisting me on two research trips to Florida and for giving me access to their wonderful collection of Carl Fisher photos. Jill Severn at the Richard B. Russell Library for Political Research and Studies at UGA helped me with every stage of this project and even tracked down photos for me on her own time. Even after crippling state-budget cuts reduced their staff to just a few, the exceptional archivists at the Georgia Department of Archives and History were generous with their time and assistance. I am especially grateful to Steven Engerrand for his help during my last couple of research trips to Atlanta and for facilitating the process of obtaining permission to use their maps and photos. Gary Doster and Ed Jackson gave me access to their impressive private collections of Dixie Highway photos, two of which appear on the cover of this book. The kind folks at the Whitfield Murray Historical Society in Dalton, Georgia, not only loaned me one of my favorite images in this book but also invited me to give a talk at a particularly critical time during the revision process. Their feedback helped me to rethink the overarching point of this book, and it is all the better for it. I am grateful to them all, but especially to Jennifer Detweiler and to Judy Alderman, a gracious lady and fine historian whose story about the origins of the term "peacock alley" is a lot better than mine, and no less true.

I'm fortunate to have had Chuck Grench at UNC Press as my editor. I am thankful for the time and effort that he, Sara Jo Cohen, Allie Shay, Lucas Church, and especially Jay Mazzocchi put into this book. My thanks as well to Sian Hunter, who supported this project from the very beginning. Above all, I'm indebted to my anonymous readers for UNC Press, whose encouragement and constructive criticism helped me to sharpen my arguments and polish my writing.

As I began revising this manuscript, I spent three wonderful years as the Kirk Visiting Assistant Professor of History at Agnes Scott College, where I received substantial support for my research and writing. Thanks to Kathy Kennedy, Mary Cain, Violet Johnson, and Shu-Chin Wu for making me feel so welcome there. A year as a postdoctoral fellow at the Center for the Study of the American South at UNC–Chapel Hill allowed me to complete the major revisions on this manuscript, and I am grateful to Harry Watson and Sally Greene for making that year such a productive one. Funding from the history department, the dean of the School of Humanities and Social Sciences, and the college-wide research and development committee at the College of Charleston allowed me to finish the book. I thank my colleagues at C of C for their support and encouragement, as well.

Over the past decade, a group of excellent mentors and colleagues have seen me through this work. At Yale, John Mack Faragher, Matthew Jacobson, and David Blight helped to guide me through the initial stages of this project. But my friends and colleagues from graduate school and beyond have been excellent mentors, as well, especially Kat Charron, Adriane Lentz-Smith, George Trumbull, Robin Morris, Claire Nee Nelson, Wendy Warren, Erika Stevens, Scott Poole, Jason Coy, Lisa Pinley Covert, and Tanya Boggs. Special thanks to Jim Giesen for giving me a writing playlist and a pep talk during the final throes of revisions.

Angela Pulley Hudson and I began our academic careers together at UGA and then Yale and both ended up writing about roads. This surprised us both but shouldn't have, since we hammered out the basic ideas behind our dissertations over dozens of happy hours in New Haven. I hold Angie up as not only a scholarly example but also a personal one. I'm proud to be her colleague and friend, and I am thankful that she is part of my extended family.

I admire Honor Sachs for her humor and intelligence, and I am heartened by her support and generosity. She has read every version of every chapter of this book at least once and offered extensive feedback, even while working on two books of her own. Honor's ability to see this project in its entirety when all I could see were the corners I'd written myself into is the main reason this book is finished. I am so lucky that she is my best friend, my editor, and my sounding board because all three require superhuman levels of patience.

My adviser at Yale, Glenda Gilmore, is the best in the business. Even after I left graduate school, she worked with me on this manuscript with the same energy and optimism and reminded me time and again why writing history is important. Glenda gave me an intellectual home in New Haven, but she also welcomed me into a circle of expatriates who remain among my closest friends. She makes a mean late-night burrito, too.

My family had little to do with the process of researching and writing this book, but I owe them the most gratitude of all. My four-legged kin, Maggiebeast and Clyde Barrow, have slowed down the pace of my work with their incessant demands for walks, treats, and belly rubs, but they are excellent company. My mother never asks me about my work, but she always asks me about me. My sister Amy and brother-in-law John expanded our family in June 2010 with a little nugget named Shiloh who gives me all kinds of hope for the future. My beloved daddy lost his battle with pancreatic cancer just three months later and before this book was complete, but he would have been proud of it like he was proud of everything I did, whether I deserved it or not. This book is dedicated to my mama and to him for teaching me to navigate those back roads, among so many other things.

Dixie Highway

Introduction

This is a history of the Dixie Highway, a hugely ambitious route built between 1915 and 1926 that proved the promise of the automobile age and helped inspire a federal highway program. Made up of hundreds of short, rough, local roads stitched together into a continuous route, the Dixie Highway looped nearly 6,000 miles from Lake Michigan all the way to Miami Beach and back again. It was originally conceived as a single tourist road to steer wealthy motorists from cities such as Chicago and Indianapolis through the South on their way to fancy vacation resorts in south Florida. Yet within a few short years, the Dixie Highway became a full-fledged interstate highway system—the first in the country's history—and served tourists, businessmen, farmers, and everyday travelers alike. By eroding distinctions between old farm-to-market roads and new automobile tourist highways, the Dixie Highway galvanized broad public support for modern state and federally funded roads and highways in the twentieth century.

The life span of the Dixie Highway was brief but exceptional. It began as an experiment by auto industry pioneers and their allies in the Progressive Era Good Roads Movement, a loose confederation of individuals and organizations committed to improving the nation's roadways. When the route was first proposed in 1914, the only good roads in the nation were in the urban Northeast, where denser populations, shorter distances, and market necessities had produced fine city streets and passable intercity routes in the nineteenth century. Elsewhere over the vast continent, atrocious roads administered by county officials and inadequately maintained by convicts or statute labor stifled the economy and isolated Americans from one another. The southern United States, increasingly populated and nurturing nascent industry, found itself imprisoned by often impassible roads that linked farms to only the nearest railroad depot. By bridging North and South, the Dixie Highway promised to both end the region's isolation and serve as a model for modern long-distance automobile routes. In many ways, it was successful. By the mid-1920s, it was the backbone of thousands of miles of new and integrated state and federal highway systems. Although it soon faded from memory, the Dixie Highway left an indelible mark on the modern highway system.

The Dixie Highway served as a model not only for highway reform but for political reform as well. Building public thoroughfares, even ones planned and administered by private organizations like the Dixie Highway Association, required financial and administrative resources beyond the means of most local road commissioners. In the Dixie Highway's brief lifetime, road construction and maintenance passed from the sole jurisdiction of local officials into the hands of state and federal highway experts. And what began as a project to build an interregional tourist route exploded into a national dialogue over the politics of state power, the role of business in government, and the influence of ordinary citizens.

In the South, where both roads and politics served to isolate the region from the rest of the nation, these transformations were the most pronounced and consequential. This book argues that road building was a crucial linchpin in the transition to the modern South, a transition that shaped the region's political institutions as much as its infrastructure. As the nation began to shed its nineteenth-century past—and with it, a unilateral dependence on railroads for long-distance transportation—road building propelled the country, and especially the South, into the modern age. As the first major interstate route to bridge North and South during a time when most roads were built by local governments for local use, the Dixie Highway was far more than just a road. It symbolized the possibilities and limitations of the American can-do spirit in an increasingly complex world. Its very existence both inspired and reflected the sweeping changes under way in the South and the nation.

This book also challenges the prevailing assumption that southerners, who were historically suspicious of federal government intervention and loathe to pay for public works projects (which often disproportionately depended upon farmers' property taxes), automatically eschewed "big government." During the Progressive Era, they recruited it, shaped it, and enjoyed its fruits. Nowhere was federal intervention more conspicuous, or southern interest in it greater, than in road building. This book explores that process by showing how southerners linked hands with midwestern automobile men in the Good Roads Movement and lobbied government bureaucrats to build the modern roads and highways that county governments could not build.

Road building played a decisive role in the transition to the modern South, but that transition occurred with a series of twists and turns. As the power required to build long-distance highways became concentrated in the hands of state and federal officials, the decision-making power of local people was diminished and a populist backlash arose. This response to federalization, coupled with the onset of the Great Depression and then World War II, stalled significant progress in road building in the South for the next three decades. Ultimately, it took a soldier who had witnessed the Dixie Highway experiment while stationed in Georgia during World War I to revive the highway program in the 1950s. Dwight D. Eisenhower never forgot the lessons that the wartime highway crusade taught him. The interstate highway system is his legacy, but in a way it is the Dixie Highway's legacy as well.

■ ■ ■

Despite its importance, the Dixie Highway has been largely forgotten in popular memory and all but ignored by historians, except as an anecdote related to early auto tourism and the origins of roadside architecture.[1] Indeed, very little at all has been written about roads in the first quarter of the twentieth century. A few writers have examined the New York–to–San Francisco Lincoln Highway, initiated just a year before the Dixie Highway, but none have engaged the larger political or social complexities of road building in that era.[2] In contrast to the scarcity of historical studies of early road building stands a wealth of fine scholarship on southern Progressivism and the emergence of the modern South. Yet, despite their significant impact on the political and economic developments of that period, roads are not addressed in any depth in these works.[3]

Although the stakes were higher for the Good Roads Movement than for other policy changes or reform movements that required fewer resources and less political reorganization, it too has received remarkably little scholarly attention. A few studies have explored the supposed tensions between the urban bicyclists who founded the Good Roads Movement and the farmers who later joined it. A handful of good articles focus on local road-improvement efforts in rural areas of the South and West. These works comprise a small but fine body of scholarship, but none link the history of road building during that era to the larger significance of the Progressive reform agenda.[4]

Howard Preston's excellent *Dirt Roads to Dixie: Accessibility and Modernization in the South, 1885–1935* is a notable exception, but it diverges from *Dixie Highway* in important ways. Preston argues that good roads "lost their significance as a reform issue" after the Good Roads Movement was taken over by New South boosters such as John Asa Rountree, "wolves in sheep's clothing" who promoted tourist highways at the expense of farm-to-market routes. *Dixie Highway* argues against this traditional, divisive view of road building and shows that farmers and businessmen, northerners and southerners were for many years united in their support of the Good Roads Movement. This is key to understanding not only how such an ambitious interregional project like the Dixie Highway was completed but also why bold, expensive new state and federal highway legislation proliferated during the Progressive Era. And while Preston explores the ways numerous new roads facilitated modernization and replaced "the region's cultural identity with a wholesale, predictable sameness," *Dixie Highway* eschews an emphasis on sweeping cultural changes in favor of a close look at the political and social consequences of modern highway construction, using the story of the Dixie Highway—which Preston addresses only briefly—to explore the complicated interactions of local, state, and federal highway agencies and ordinary citizens.[5]

The overwhelming majority of historians who have written about the highway revolution have focused on the Eisenhower interstate system of the 1950s. Many others have written about the early automobiles that fueled the Good Roads Movement but have ignored the dirt roads on which those cars ran—or tried to run. While both of these topics are important to the history of American transportation and southern modernization, the heavy focus on them has obscured the central role that roads played in some of the most important political debates and reform movements of the early twentieth century.[6]

Dixie Highway therefore fills a significant void in the historiographies of southern Progressivism, modernization, and transportation. By using one of the most successful Good Roads Movement projects as a narrative device, this book illuminates the debates that shaped Progressive Era politics in the South as well as the development of the modern highway system. It seeks to restore the Dixie Highway and the politics of road building to the history of the Progressive Era and the history of the modern South. These linkages are fundamental to our understanding of the modern South, for roads both shaped and reflected the development of the modern South during a period of tremendous growth and change.

Roads were catalysts for a chain reaction of economic developments, but they depended heavily upon new modes of transportation. Automobiles—arguably the most visible symbol of twentieth-century modernization—were especially popular in the South and intensified the demand for better roads. Cars and trucks, in turn, facilitated not only individual travel but economic growth and diversification as well. Southern manufacturing in the 1910s and 1920s expanded as new and better roads supplemented and in rare cases replaced railroads. For example, the southern textile industry, which was first introduced in the region in the 1880s and by the 1920s had become the world leader in cloth and yarn production, depended upon local roads. Historians rightly acknowledge the important role railroads played in establishing the textile industry in the South after the Civil War and linking it to markets nationwide, but new and existing roads provided essential links between fields and factories.[7] Most local road systems resembled spokes on a wheel, with roads branching out into the countryside from the town railroad depot.[8] These crude networks of dirt roads delivered countless bales of cotton to boxcars bound for the burgeoning textile-mill cities in piedmont North Carolina, upcountry South Carolina, and north Georgia.[9]

Modern routes like the Dixie Highway challenged both the old spokes-on-a-wheel model of road building as well as the railroad monopoly on long-distance travel and trade. However, while most southern industries remained dependent upon railroads for many more years to come, farmers did not. Agricultural diversification, the cornerstone of the U.S. Department of Agriculture's extension work by the 1910s, depended upon easy access to more-distant markets. As farmers shifted from cash-crop production to new crops such as peanuts, soybeans, and corn, most rural railroad depots still linked farmers only to markets for cotton and tobacco. Farmers and county extension agents viewed the good-roads campaign as the preeminent partner of the diversification program. In fact, the Bureau of Public Roads, which coordinated the earliest roadwork by the federal government, was a subagency of the U.S. Department of Agriculture (USDA), and early federal highway policy focused exclusively on rural roads. Local roads and long-distance highways also facilitated the industrialization of agriculture by enabling long-haul trucking, a more flexible and affordable alternative to railroads.[10]

Roads transformed the South by introducing new goods and services to the region as well. The introduction of Rural Free Delivery in the 1890s quickly made the mail-order catalog business a staple of rural households.[11] Automobile tourism transformed cities and rural hamlets alike by generating new business opportunities. Gas stations, motels, and restaurants soon lined the streets of cities and towns, while roadside stands and auto camps dotted rural landscapes along new routes like the Dixie Highway.[12] In the north Georgia town of Dalton, enterprising women selling hand-tufted chenille bedspreads to Dixie Highway travelers spawned a whole new industry. Historian Douglas Flamming argues that Dalton's bedspread trade was the most significant economic development in north Georgia during the interwar years and continued to play a major role in the local economy for decades thereafter.[13]

Roads and highways also influenced political and social institutions. Indeed, much of the Progressive Era reform agenda in the South was inextricably linked to roads. Alongside white-supremacist reforms such as African American disfranchisement, county chain gangs supposedly rehabilitated prisoners once exploited by the brutal convict lease system and improved public roads more efficiently and economically than the older systems of volunteer or statute labor. Advocates of rural school consolidation and compulsory attendance, two cornerstones of Progressive reform in the South, recognized the connections between good roads and modern, more accessible schools.[14] Even Prohibition—both enforcing it and violating it—depended on good roads as much as fast cars. The Dixie Highway earned the nicknames "Avenue de Booze" and "Rummers' Runway" because it was often used to transport alcohol from Canada—just across the narrow St. Marys River from the highway's northern terminus of Sault Ste. Marie, Michigan—to distribution points throughout the Midwest and the South. Federal agents prowled the route as well, often coming to the aid of local authorities overwhelmed by the "Great Booze Rush" on the Dixie Highway.[15]

The Dixie Highway offers the best case study of the links between Progressive reform and road building because it was so exceptional. It was at once a local, state, and interstate route and an interstate system, with parallel north-south routes connected by short east-west roads. Unlike most other named routes (or "marked trails," as they were commonly called), the Dixie Highway was actually completed. The Lincoln Highway, the Dixie's most famous contemporary, was not as complex a route as the Dixie, and parts of it remained unfinished by the time dozens of numbered state and federal highways absorbed it in the mid-1920s. The Dixie Highway was the only major thoroughfare to crisscross the South, and it inspired countless feeder roads to link isolated citizens to distant cities and markets. And because its origins and construction overlapped with significant developments in state and federal highway legislation during the 1910s and 1920s, the Dixie Highway offers an ideal lens for examining the associations between modernization and Progressive policy making in the South.

The Dixie Highway may have owed its success to the enthusiasm of southern good-roads devotees, but it was actually the brainchild of Carl Graham Fisher, an opportunistic auto-industry magnate and real-estate developer from Indianapolis. As they became more popular in the 1910s, automobiles created new challenges for primitive local roads that entrepreneurs like Fisher were eager to exploit. They found eager partners in the already-thriving Good Roads Movement. Frustrated with the slow pace of local road construction, both ordinary citizens and auto industrialists like Fisher lobbied for greater government investment in highways. They called for levels of state and federal aid akin to that of the massive railroad projects of the nineteenth century, which they hoped modern highways would soon surpass as the major arteries of commerce and travel. The Dixie Highway was a blueprint for the kinds of modern roads they envisioned.

But many never quite imagined the ultimate outcome of their project: a government bureaucracy that managed roads in a way that marginalized citizen input and supplanted the coalition forged by the Good Roads Movement. Thanks to the efforts of the Dixie Highway's tireless supporters, the several new state highway systems along its path and, finally, the first numbered system of U.S. highways absorbed the Dixie's route. A multitude of numbered highway signs replaced the familiar red and white "DH" markers that had guided motorists along the bumpy roads for more than a decade, erasing any traces of the Dixie's original path. By 1926 it had disappeared from the map, a victim of its own success. Its amateur organizers, more skilled in boosterism than in planning and engineering, found themselves replaced by trained professionals, a familiar story in the Progressive Era. Citizens who supported the Dixie found their influence weakened, as well, as highway bureaucrats outranked local elected officials in designating and funding roads.

In the South, the backlash against the modern highway bureaucracy embodied the very essence of southern Progressivism by highlighting the limits of social and political reform. Southerners embraced state and federal aid, but only when they could mold it to fit their needs. As long as the consolidation of control over road building did not threaten cherished southern institutions—most notably racial control, as evidenced by the widespread support for predominantly black prison chain gangs—white southern voters embraced state and federal intervention. But when bureaucracy threatened those institutions and shut out local input, southerners retreated from supporting strong state and federal highway programs. Their response forecasted resistance to federalization during the New Deal and helps to explain why it took thirty years before southerners again embraced the modern highway bureaucracy.

■ ■ ■

Although it is difficult to imagine today, when easy access to multilane interstate highways and elaborate grids of city streets are taken for granted, in the 1910s and 1920s, a crisis of scarce roads dominated political debates, business schemes, and a major social reform movement. The Progressive Era Good Roads Movement was one of the nation's most visible campaigns for social, economic, and political integration, and the Dixie Highway was its most profound accomplishment.

This book addresses questions central to understanding the reform impulse of the Progressive Era. How did a modernizing project like a highway fit into reform agendas? And what were the limits of reform? By 1910 the United States was the world's leading industrial power, a global symbol of wealth and progress. Rapid technological changes, an influx of immigrants, and unprecedented urbanization exposed social problems that a wide array of reformers attempted to address. The Good Roads Movement, which began among urban bicycling enthusiasts in the 1880s and spread to rural areas as the popularity of automobiles grew, exemplified Progressive efforts to solve problems exacerbated by technological advances and population increases. As many historians have argued, another important dimension of Progressive reform was the emphasis on social order. In the South, African American disfranchisement, prison labor, and new ideas about state and federal intervention characterized Progressive reform.[16]

Along with other major infrastructure projects of the time, including the Panama Canal and the Golden Gate Bridge, the Dixie Highway resonated with both of these facets of Progressivism.[17] Certainly it tackled the nation's transportation problems using both engineering techniques and a new bureaucratic government apparatus, but in the South, it also conformed to efforts to control black labor and preserve local political control. Opportunistic auto industrialists embraced these institutions in order to get what they wanted as well. The Progressive Era Good Roads Movement gave both automakers and farmers a place to formulate and express their own limited, self-serving visions of reform.

Capitalists and citizens, northerners and southerners, and city dwellers and farmers all lived and worked along the Dixie Highway and joined together in a united effort to complete the nation's first interstate highway system by securing state and federal funding. Guided by the businessmen who formed the Dixie Highway Association, they used the Good Roads Movement to keep transportation issues at the forefront of local, state, and federal politics. Their campaign opened up new markets for the auto industry in the South at the same time it provided unprecedented business and marketing opportunities for road-deprived rural southerners. The Dixie Highway experiment revealed striking differences between the ways that northerners and southerners envisioned road building, as southerners proved committed to the use of chain-gang labor even as northern states committed themselves to more aggressive forms of funding to pay for engineers, wage laborers, and modern machinery. Yet for all their differences, businessmen and farmers faced a common need—transportation—and forged a coalition to satisfy it.

What no one in the Good Roads Movement envisioned were the kinds of struggles that embroiled the Dixie Highway in political contests over the role and place of government investment in local affairs. The creation and enlargement of state and federal bureaucracies threatened local control, especially in the Deep South, where the tension between the desire for big roads and the limits of big government were most pronounced. Coupled with resistance to federalization during the New Deal, this tension crippled the Good Roads Movement and compromised roadwork for years to come.

■ ■ ■

This book examines the Dixie Highway in several states, but it focuses on Georgia, where the juxtaposition of a weak, conservative central government and an extensive transportation network threw the links between road building and state building into sharp relief. One of the largest states on the entire Dixie Highway route and the largest in the South, Georgia also had more Dixie Highway mileage than most other states on the route and also was the gateway to Florida, the destination for the wealthiest of the Dixie Highway's motorists. Georgia's roads were of utmost importance to the businessmen and boosters behind the project and thus figure prominently in reports of the Dixie's progress. The parallel struggles of those men and ordinary rural Georgians to expand and improve roads shed light on the conflict that eventually fractured the Good Roads Movement.

The populist backlash against the expanding highway bureaucracy stalled the highway revolution for another two generations. Not until after the Second World War, when the nation was stronger and wealthier than ever before, would the United States finally be ready to face the scale of federal investment that modern interconnected highways required. But when Americans did turn their attention to building highways, their vision for this massive undertaking would not emerge out of thin air. The template for a government-funded interstate highway system had roots in the Dixie Highway. In the 1910s and 1920s, the Dixie Highway carved inroads—both literally and figuratively—into the nation's ideological and geographic landscapes. While contemporary observers became embroiled in the political conflict of roads, the Dixie Highway opened up new possibilities that would ultimately become reality a few generations later.

Nowhere is this connection clearer than in the figure of Dwight Eisenhower, whose name will forever be closely associated with the modern interstate highway system. Eisenhower not only witnessed early auto experiments along the Dixie Highway during World War I, but he also participated in similar exercises after the war. The experiments that Eisenhower witnessed as a young man would have an enormous impact on his future policies. Eisenhower saw the seeds of possibility as a young soldier watching convoys make their way along bumpy, rutted roads. He took to heart the struggles surrounding the origins of systematic highway management and the nation's inexperience with it in the early twentieth century. As the country sought to understand the full implications of the new era, Americans grappled with everything that a road represented—expanding government, modern technology, car culture, and the demise of the imagined social order enshrined in an idealized past.

But to a young observer like Eisenhower, a road represented much more. It represented a vision of modern routes that knit the nation together through a fully conceived, federally funded interstate highway system. The Dixie Highway sparked debates about national roads, and it left an ideological and political legacy that would later be realized on a grand scale.

This book explains the origins of that legacy. Chapter 1 contextualizes the rough state of transportation that existed before 1915, when muddy local roads ensnared travelers, isolated entire communities, and frustrated early proponents of the Good Roads Movement. It describes the rise of the auto industry alongside the struggles of southern farmers and explains how the road crisis forced both to acknowledge the need for a larger state and federal role in highway work. Placing farmers and automakers squarely within the context of Progressive Era road building and state building repositions them as the central reformers in the Good Roads Movement and the architects of named highways and marked trails like the Dixie Highway.

Chapters 2 and 3 follow the Dixie Highway Association and its supporters as they launched a successful national campaign to simultaneously build an interstate highway and lobby for state and federal aid with which to complete it. The routing competition for the Dixie Highway in the winter and spring of 1915 described in chapter 2 demonstrates how the Dixie Highway Association forged a broad-based coalition of automakers, civic boosters, local road officials, and ordinary citizens to shape the contours of the highway. The outcome of the fevered contest to determine the exact route inspired Dixie Highway officials to augment their plans for a single route and build an interconnected interstate highway system instead. But the routing contest also exposed significant regional divisions between the road-building capabilities of local governments, while the outcome of the contest multiplied the already considerable challenges of completing the highway. This, in turn, intensified the efforts of Dixie Highway supporters to secure state and federal funding for their ambitious experiment. Chapter 3 tracks their efforts up close by following the Dixie Highway Association's World War I campaign to secure government support for roads. By recasting through routes like the Dixie as military necessities, they persuaded elected officials to rethink the limits of state and federal intervention in public works.

Chapter 4 traces the limits of the Dixie Highway Association's successful road- and state-building campaigns in the South. The rise of the chain-gang road-labor system produced mixed results for the highway lobby's crusade to secure more state and federal highway revenues. As the Dixie Highway moved forward, striking differences emerged between the way northerners and southerners continued their construction operations. At the heart of these differences was the South's unwillingness to surrender local control over convict labor and, ultimately, white supremacy, in spite of the costs to the modern highway-building campaign.

The final chapter follows the final years of the Dixie Highway experiment, when fears over corruption and too much state power derailed the Good Roads Movement. Starting with the contentious routing contest that erupted after the announcement of the first numbered U.S. highway system in 1924, southerners began to sour on the demands of modern road building. As state and federal highway agencies began to replace named highways with the kinds of integrated state and federal routes that good-roads advocates had always wanted, southerners grew wary of the magnitude of the task and the faceless bureaucracy it required. Such wariness came to fruition in the contentious Georgia gubernatorial race of 1926, which pitted a state highway bureaucrat against an opportunistic politician eager to exploit populist anxieties about government power. As road building emerged at the center of political debate in 1926, it embodied the tensions between local desire for modern highways and discontent with the price, both financial and political, of building them. Those tensions are still with us today.

Building a Good Roads Movement, 1900–1913

At the turn of the twentieth century, a host of humorous songs, poems, and anecdotes poked fun at the nation's abysmal roads. A poem called "Bad Roads Did It" played with the common theme of farmers getting stuck in the mud:

> A farmer old, so we've been told,
> With a team of horses strong
> Drove down the road with a heavy load,
> While singing his merry song.
> But his mirth in song was not so long
> For his horse gave a leap
> As he ran amuck in the mud he stuck,
> Clear up to his axles deep.
> Bad roads did it![1]

Another example relayed the tale of two farmers who could only communicate with one another across the muddy, impassable road that separated their two farms. While talking one day, they noticed a hat lying atop a mud puddle in the middle of the road. It was a very fine-looking hat, so one of the men made up his mind to retrieve it using a fishing rod. He cast his line into the mud and managed to hook the hat. As he began to reel it in, however, a man's head suddenly emerged from the mud beneath the hat. "Why you poor fellow," the fisherman exclaimed, "you are in an awful fix!" "Well, I don't mind it very much," the man calmly replied, "but it is pretty hard on the horse."[2]

Comical stories like these mocked bad roads, but they constituted part of a very serious national dialogue about the limits of transportation options. In the early twentieth century, roads governed everyday life. They determined where people could and could not travel, as well as whether or not other people, goods, services, and even ideas could reach them. Roads dominated conversations around the ballot box and the dinner table, but good roads eluded most Americans and virtually all southerners. In their place was a jumble of muddy sand-clay routes covering the region like a bed of briars, sometimes radiating from a railroad depot but often simply from each other, full of dead ends and treacherous mud puddles like the one that snared the apocryphal traveler and his horse.[3]

The bureaucracies that managed roads were equally underdeveloped. Farmers and other citizens who depended on these routes had no one to turn to for help besides the local officials who were already failing to maintain them. Local government control over road building divided scarce resources and limited opportunities for travel and trade. There were few links between towns, and even fewer between states and regions, because no central agency coordinated roadwork across local or state boundaries. Only seven state highway departments existed in 1900— none in the South—and no federal agency was empowered to fund road improvements.[4] No one saw much need for long-distance roads, anyway. Horses and wagons were not long-distance modes of transportation for the average traveler, so aside from a handful of old stagecoach routes and Native American trading paths, roads linked farms to the nearest towns and to little else.[5] Roads were entirely local in scope and in management, and they circumscribed people's lives accordingly.

Railroads crisscrossed the South by 1900, but they were no substitute for a good system of roads. While railroad lines reached deep into rural communities, most had monopolies over local trade and travel. These monopolies resulted in high freight rates and limited passenger schedules that severely limited ordinary people's ability to get around.[6] Railroads did not ease the isolation or the financial pressures of farmers who needed to travel where railroad lines did not go. When bad weather or general neglect made roads leading to and from the depots impassable, railroads were of little use. With the average farm ten miles from a depot, rural families at the turn of the century were at the mercy of a woefully deficient system of local dirt roads.[7]

The invention of automobiles around the turn of the century challenged the limitations of local roads. While initially they aggravated tensions between wealthy urban motorists and the farmers whose roads they carved up on weekend drives, cars became more common in rural areas of the South after 1910.[8] The range and flexibility of automobiles required bigger, better roads, which neither urban motorists nor rural farmers alone could afford to build and whose construction would have to depend on the coordination of dozens of county and city governments. People had to rethink not only the way roads looked but also how they were built and maintained. Over the next few years, cars eroded geographic, socioeconomic, and political divides. Farmers and motorists alike came to view roads not as farm-to-market routes or automobile thoroughfares but as parts of a modern state and federally built highway system that served a range of different interests.

By recognizing their common interests, rural farmers, city drivers, and automobile industrialists forged an uneasy alliance in the Good Roads Movement in the 1910s and 1920s. Such wide-ranging support was not altogether unusual for a Progressive reform campaign. For example, Prohibition supporters included women seeking political leverage for a future suffrage amendment, religious leaders concerned about lax morals, and "wet-dry" politicians who drank openly but jumped on the dry bandwagon when it was politically expedient. Those who backed electoral reforms included advocates of everything from secret ballots to African American disfranchisement.[9] Still, the alliance between urban northerners and rural southerners, and between industrialists and farmers, was exceptional. Divided by political and economic priorities, not to mention sectional divisions reinforced by anemic lines of communication and transportation, this motley mix of people came together to challenge the status quo in road building.

The automobile served as the catalyst for farmers' and auto industrialists' joint efforts to redefine both good roads and good government. It was out of this coalition that the Dixie Highway was born. By 1915, these road advocates envisioned long-distance automobile highways like the Dixie Highway as the backbones of new, modern highway systems that connected isolated farms with distant metropolises. The campaign pushed the limits of automotive technology, challenged the hegemony of the powerful railroads, and forced both automakers and their fastest-growing group of consumers to compromise on some of their most sacred beliefs about the federal government's proper size and reach. In the new century, big roads, big business, and big government remade the Good Roads Movement and the South alike.

The Origins of Good Roads

In the early 1900s, local governments had neither the resources nor the expertise to manage roadwork. Building just one mile of paved road cost an average of $5,000 in 1909, and fancier paving materials were twice that much. A mile of "improved" dirt road—meaning graded and surfaced with a sand-clay mixture that was supposed to hold up better than topsoil but rarely did—was a fraction of that cost at $723 per mile, but even small counties had hundreds of miles of rough roads linking farms to towns and could scarcely afford to improve them all. Northern, urban states had the nation's best roads, but even there, the figures were grim. In New York, barely 16 percent of the state's roads had been improved by 1909, and New Jersey, a much smaller state, had improved less than a quarter of its roads (and most of those were concentrated in cities). Massachusetts and tiny Rhode Island boasted nearly 50 percent improved roads. Roads in the rapidly urbanizing Midwest were good, too. Ohio had improved over 27 percent of its roads by 1909, down from over 30 percent a few years before, most likely because of population increases and greater wear and tear. Nearly 40 percent of Indiana's roads were improved. The percentage in Illinois was much lower, but urban counties averaged between 50 and 80 percent. Moving South, where states were larger and tax revenues smaller, the numbers dropped considerably. Kentucky had improved only 18 percent of its roads by 1909, and Tennessee had improved scarcely 11.5 percent. Georgia, the largest state in the South, had improved just over 7 percent of its roads. Only the larger but more sparsely populated states of the West fared worse.[10]

Rural southern counties struggled to keep up with the costs of maintaining the roads they already had. Due to higher tax values and larger local road bonds, the average expenditure for roads per resident in the Northeast in 1910 was $1.53. In the South, it was $0.86. Denser populations along northern roads further increased the funds available to improve each mile of road. In the South, where population densities were half that of New England and 80 percent that of the Midwest, counties had to find alternative ways to finance roadwork.[11] Before local and state laws were passed to put convicts to work on public roads, many southern counties relied upon statute labor laws, which compelled able-bodied male citizens to donate their time—and their tools—to work on local roads a certain number of days each year. Enforcing these unpopular laws proved impossible in many counties, while in others, men often paid a minimal tax in order to get out of working.[12] In Chattooga County, Georgia, county road supervisors reported that as many as half the eligible men in some of the county's ten districts paid from one to two dollars in order to get out of working the roads. The money presumably went back to the county road fund, though Chattooga officials did not spend it to hire additional labor. Even when men did "work out" the tax instead of paying it, they accomplished little: in Chattooga County districts, the average work crew consisted of approximately ten men working a specific stretch of road with shovels and a paid "overseer."[13] Since most crews in Georgia were required to work only about five days out of the year, it is no wonder why rural roads saw little improvement under the statute labor system.[14]

Farmers wanted better roads but felt overburdened by the higher taxes needed to pay for them.[15] But the "mud tax," as farmers called it, burdened farmers as well by multiplying the cost of transporting crops to market. A heavy wagon carrying a load of goods could easily overturn on a bumpy dirt road, spill its contents, and even severely injure the farmer driving it, or it could become mired in the deep ruts left by previous wagon wheels, which turned into mud holes every time it rained. Around 1910, good-roads disciples estimated losses from such accidents to be as high as $500–600 million per year, costs borne primarily by farmers but also by urban consumers of farm products.[16] Yet neither farmers nor urbanites were convinced they could offset the mud tax by paying higher property taxes.

The commitment to local political control compounded problems. Throughout the South, white Democratic hegemony depended upon an array of mechanisms designed to guard local control and limit state and especially federal interference in local affairs. Consequently, while many northeastern and midwestern states had developed at least minimal forms of state highway aid as early as 1911, southern states had not. They relied on property taxes, forced labor, and occasional local bond issues to finance all roadwork. And while local road expenditures in many southern states exceeded the sums issued by northern and midwestern state highway commissions, they were scattered among so many different local road projects that they had little benefit.[17] Counties and municipalities controlled only their own domains and rarely cooperated with neighboring communities on road projects. This complicated efforts to build uniform, high-quality roads that spanned multiple counties, much less entire regions. Instead, local roads often ended abruptly at farmhouses, county lines, or, more often, railroad depots, leaving farmers dependent upon avaricious railroads to get their crops to distant markets. These "spokes-on-a-wheel" systems fractured rural landscapes into a series of adjacent yet isolated communities.[18]

In Georgia, a unique electoral process known as the county unit system further protected rural control even as the state became more urban and industrial. Instituted in 1898, the county unit system mitigated the popular vote in primary elections by dividing counties into three types according to population and then apportioning unit votes for each type of county. The state's eight urban counties each received six unit votes; thirty counties with middling-sized towns each received four unit votes; and the state's 121 rural counties each received two unit votes. The candidate who received a plurality of votes in a county received that county's unit allotment. Rural votes weighed more under this system; three rural counties with a combined population far less than that of a single urban county received the same number of unit votes.[19] Residents of large cities like Atlanta, where urban issues demanded attention, resented the tyranny of their rural neighbors, but to no avail. The state's conservative leaders depended on the county unit system for their political survival and defended it vigorously. As future Georgia governor Hugh M. Dorsey proclaimed in 1914, the county unit system "is at the base of our government . . . [and] it means everything politically and governmentally to the smaller counties in Georgia. . . . [It] runs through the whole structure of our governmental scheme."[20] Thus the specter of rural political domination hung over the state throughout the years when good-roads advocates nationwide struggled to unite town and country and region and nation.

At the federal level, the idea of investing in roads was not unheard of and in fact had deep historical roots. A century earlier, Federalists and Democratic-Republicans had clashed over the limits of federal power to fund internal improvements. But as the nation expanded westward in the first half of the nineteenth century, even those most committed to small government began to tolerate greater federal responsibility for road construction, especially for post roads and military routes. Thomas Jefferson had once complained that the Federalist proposal to fund post roads "would open a bottomless abyss for public money," but after the Louisiana Purchase of 1803 doubled the size of the United States, his administration proposed a coordinated system of federal roads and canals to link the new territory to the East.[21] Fellow Democratic-Republicans, such as Henry Clay of Kentucky, supported interstate routes to connect their landlocked states to lucrative waterways and ports, while John C. Calhoun of South Carolina and other outspoken critics of federal power argued that poorer southern states deserved federal aid for internal improvements.[22] Although Congress declined to act on most road-building requests, by 1811, 37,000 miles of federally funded post roads—including the famous Cumberland Road near the nation's capital and the Federal Road between Athens, Georgia, and New Orleans—connected the East with sparsely settled areas of the South and West.[23]

Nearing midcentury, however, that progress came to a standstill when railroad fever eclipsed road building. Convinced that railroads were the routes of the future and buoyed by the consensus that indirect federal aid to railroad companies did not violate even the strictest interpretation of its constitutional powers, Congress withdrew funding for post roads and began issuing huge land grants to companies building lines westward, thereby shifting control over the nation's most important internal improvements from the federal government to privately owned railroad companies. After short-lived attempts by toll companies to maintain existing routes and build new ones, responsibility for roadwork fell entirely to local governments. Meanwhile, federal land grants to railroads enabled railroad mileage to soar from 9,021 miles in 1850 to 193,346 miles in 1900.[24] Channeling federal aid through land grants allowed the government to remain, in the words of historian Brian Balogh, "out of sight," thereby preserving the popular notion of small, fiscally conservative government while subsidizing the construction of a massive, privately owned transportation system to the tune of $500 million.[25]

Congress's love affair with railroads lasted longer than the public's. Ordinary citizens grew increasingly angry over the railroad's stranglehold on overland transportation. By 1880, railroad mileage had already exceeded that of surfaced, travelable roads.[26] Everyone suffered from the disparity, but no one more than the farmer. Major railroads had monopolies in rural areas, so they could charge steep rates for short hauls to neighboring market towns while charging lower rates on long hauls for customers in more-competitive urban markets. Because most main roads linked farms to a central railroad depot rather than to other nearby market towns, farmers had no alternatives for marketing their crops. Railroads did everything they could to preserve this system, even giving direct aid to counties to upgrade these feeder routes.[27]

The hostility that brewed between farmers and the federally backed railroads sowed deep reserves of resentment and suspicion that would haunt infrastructure projects for decades to come. In the late nineteenth century, angry farmers mobilized into powerful groups like the Farmers' Alliance and later the Populist Party to fight the railroads, but farmers were no match for the railroads' wealth or political influence. The federal government had little incentive to crack down on an industry that saved it hundreds of millions of dollars in free or reduced freight rates for military and mail shipments, a deal that had been written into federal land grants years earlier. By the 1880s, railroad industry "robber barons" like Andrew Carnegie and Jay Gould had become poster children for the excesses and abuses of power made possible by the relationship between big business and "small" government. Occasionally, Congress bent to public pressure and restrained some of the most abusive railroad practices, but the government stopped short of meaningful reform. For instance, the Interstate Commerce Commission, established in 1887 ostensibly to correct varying degrees of regulation by state railroad commissions, had few regulatory powers of its own and succeeded only in protecting the railroads from further state intervention.[28]

By the turn of the twentieth century, railroads had become almost as divisive as they once had been unifying. Cities profited from railroad traffic and some smaller towns grew into bustling market centers, while farmers watched local roads deteriorate further. Rural populations declined as a result of late nineteenth-century industrialization, so farm communities had an increasingly difficult time attracting either the quality or the variety of railroad services that helped larger towns and cities thrive. In the rural South, where there was less industry and sparser populations, few railroads and even fewer good roads multiplied the distance between farm families and their urban neighbors.[29]

From this transportation crisis emerged the Good Roads Movement. It began among urban bicyclists in New England who formed the League of American Wheelmen (LAW) in 1880, but it flourished in the countryside, where the demand for roads was most pronounced and where nearly three-quarters of the nation's population still lived. "Wheelmen" recognized the opportunity to expand their campaign and made the most of it. By 1890, LAW promoted good roads for all travelers through publications such as *Good Roads* magazine, which reached a circulation of 1 million within three years. Theirs became a sacred campaign that promised a panacea for all that plagued the farmer. A popular pamphlet, *The Gospel of Good Roads: A Letter to the American Farmer*, extolled the benefits of rural road building, promising farmers "the enrichment of your slender purse and the betterment of your condition" and urging them to lobby their state and federal legislators for roads. Soon, LAW members joined forces with other good-roads advocates to form the National League for Good Roads (NLGR). Under the NLGR's leadership, the Good Roads Movement grew further into a broad-based reform movement backed by a loose collection of local, state, and regional organizations that promoted better road construction and maintenance and raised public awareness through regular meetings, automobile contests, and political lobbying.[30] The good-roads message resonated among people in the rural South, as groups like the Georgia Good Roads Association, the Chattahoochee Valley Good Roads Association, and the Appalachian Good Roads Association proliferated around the turn of the twentieth century.[31]

In stark contrast to the public response, the federal government remained cautious toward the good-roads campaign. In 1892, after months of lobbying by the NLGR for a federal highway commission, Congress established the Office of Road Inquiry (ORI) within the U.S. Department of Agriculture and appointed Roy Stone—a civil engineer, former Wheelman, and author of the NLGR's highway commission bill—to run the new agency. A far cry from the road-building commission proposed by the NLGR, the ORI had no real authority. As historian Bruce Seely has argued, it merely served as a clearinghouse for gathering and disseminating information about building and improving roads. Furthermore, Secretary of Agriculture J. Sterling Morton ordered Roy Stone to steer clear of politics, telling him that it was "not the province of this Department to seek to control or influence" that which local governments already controlled. Even congressmen who supported the new agency wanted to curb its power. As one argued, it was "the only practical way that the federal government can take part in this great movement." Stone's $10,000 budget, which Congress reduced to $8,000 in 1896, scarcely covered salaries and basic office expenses.[32]

Weakened by a lack of federal support, the ORI was easily manipulated by the powerful railroad interests, who still viewed rural roads as feeder routes for depots rather than as potential competitors for long-distance transportation. Roy Stone cooperated with the railroads by facilitating relationships between local road officials and railroad companies, which offered discounted shipping rates on road-building materials. If endorsing spokes-on-a-wheel road building bothered Stone, he did not show it. As the good-roads activist explained to Secretary Morton, having railroads deliver materials from afar worked well because "the roads to be built will generally radiate from Railway Stations, and can carry their own material" outward from the depots. Even if road-building materials were available locally, he continued, "an independent system of good roads would be required [to transport them to the building sites], which would generally serve no other purpose."[33] Stone's successor, Martin Dodge, also nurtured the agency's relationship with railroads. During Dodge's tenure as head of the ORI, renamed the Office of Public Road Inquiry (OPRI) in 1899, railroad companies sponsored Good Roads Trains that toured the country holding conventions and building short "object-lesson" road sections, all in the service of maintaining a system of roads that served railroads. At the turn of the century, railroads were the largest corporate backers of the Good Roads Movement in the country.[34]

Good-roads advocates devised another way to solicit federal road aid through the growing demand for rural mail delivery, but this too failed to get very far. The U.S. Postal Service had provided free mail delivery to city residents since 1861, but it left farmers to fetch their own mail from the closest post office if and when the roads allowed. But in the early 1890s, Rural Free Delivery (RFD) emerged as one of the centerpieces of the Populist Party's campaign to address farmers' geographical and economic isolation. Populist Party leader and Georgia congressman Thomas E. Watson introduced a bill to fund rural mail service, and it was signed into law in 1896. Almost immediately, however, support for the new law evaporated as the economic implications of building thousands of miles of new post roads became clear. The RFD bill also pitted congressmen against some 77,000 angry small-town postmasters whose federal patronage jobs would be imperiled by home mail delivery. Opposition from the postmaster general, President Grover Cleveland, fiscally conservative congressmen, and countless small-town merchants and private express companies obstructed new post-road construction for years after the RFD bill passed, thereby restricting the new service to the few areas where travelable roads already existed. In 1900 approximately 4,000 RFD routes served 3.5 million rural Americans.[35] This was less than 5 percent of the total population and only about 7 percent of the nation's rural population.[36]

While federal support for the Good Roads Movement stagnated at the turn of the twentieth century, powerful business interests began to shift the tide. RFD generated tremendous public interest from newspapers, magazines, and mail-order catalog companies eager to extend their business into country homes. It proved even more popular among rural people themselves, who saw in this new service viable connections to the outside world and to the growing corporate-consumer economy that railroads had kept largely beyond their reach. After 1900, under pressure from Progressive reformers as well as powerful lobbying groups backed by major mail-order companies such as Sears, Roebuck and Co. and Montgomery Ward, congressional support for RFD and post-road construction began to expand again.[37] The popularity of RFD helped to persuade many farmers to join good-roads clubs, especially in the South. As membership expanded, so did the number of local and regional clubs. In 1902 only a dozen good-roads groups existed in the South. Within a decade, there were 274, constituting approximately 60 percent of good-roads clubs nationwide. Georgia alone had fifty-four clubs, more than any other southern state except North Carolina.[38]

By 1910 good-roads crusaders formed a powerful voting bloc, yet they still struggled to influence far-reaching improvements in roadwork. While they had discovered that alliances between rural reformers and business interests could prove enormously fruitful when trying to gain support for the Good Roads Movement, they never fully solidified relationships among member organizations. No single leader or central organization coordinated their efforts, and no government agencies above the local level had the authority to help. Despite widespread support, the Good Roads Movement was too scattered and too weak to revolutionize the country's transportation system one county road at a time. It took the invention and popularization of the automobile to change that. After 1910, cars jump-started the modern Good Roads Movement by giving new urgency and new direction to the campaign to build roads.

Making Good Cars and Good Roads

In 1903 bicycle manufacturer-turned-automaker Colonel Albert Pope declared that "the American who buys an automobile finds himself with this great difficulty: he has nowhere to use it." Indeed, many people thought the scarcity of good roads would limit the automobile's appeal, but sales still soared.[39] First manufactured by German engineer Gottlieb Daimler in 1885, automobiles powered by steam, by electricity, and, most successfully, by gasoline were introduced in the United States in the early 1890s. The first American-made car was sold in 1896, and by 1905 U.S. automakers had surpassed Europeans in both quantity and quality of their machines. Around the same time, the center of the fledgling U.S. automobile industry moved from the Northeast to the Midwest, where gasoline engines were already being manufactured for farm use due to the lack of electricity in rural areas and nonunion labor and abundant raw materials lowered production costs. This shift, along with the superiority of internal combustion engines for navigating the Midwest's bad roads, virtually guaranteed the dominance of gasoline-powered cars over steam or electric alternatives. In 1907, 44,000 automobiles were manufactured and sold in the United States. Domestic sales skyrocketed to 65,000 when Henry Ford's affordable Model T was introduced the following year. Some 10,607 Model T cars were sold in 1909–10, and sales more than tripled to 34,528 by 1911. In 1910, 187,000 cars were made in the United States alone. Total registrations of both American- and foreign-made cars in the United States numbered 458,500 by 1910, more than any other country in the world.[40]

When cars first appeared, they threatened to fracture the Good Roads Movement along existing class and regional fault lines. Reliable cars cost at least $700–800 and often twice that much, far beyond the means of average wage earners or farmers.[41] Early automobiles were playthings for wealthy metropolitans who liked to flee the confines of the city for weekend drives. Farmers, however, detested the loud, dangerous "devil wagons" that zipped through the countryside tearing up dirt roads and frightening—and sometimes striking—livestock. Some resisted by refusing to move their horses and wagons to the side of the road in order to allow cars to pass. Others found more aggressive ways to thwart motorists, including scattering tacks and glass in the roads to puncture tires or stringing rope or barbed wire across the road. In South Carolina, one frustrated farmer even fired his pistol at a car that was attempting to pass his horse-drawn wagon.[42]

Such resentment compounded the difficulties of improving local roads by creating new tensions over which kinds of roads should get the lion's share of limited local road revenues. Farmers bitterly opposed the prospect of using their property taxes to build long-distance automobile routes, which they nicknamed "peacock alleys" after the corridors in fancy hotels where wealthy guests strutted about in their expensive clothes and jewels.[43] City residents who had already taxed themselves to build and maintain their own streets, on the other hand, wanted better country touring roads but were reluctant to finance routes that would primarily serve farmers hauling their crops to market. Both sides exaggerated the differences between the two types of roads, but only because they conceived of roads as strictly local responsibilities. A stalemate developed out of these tensions, threatening not only local efforts to improve existing roads but also the cohesiveness of the Good Roads Movement's campaign to build new roads.[44]

While automobiles magnified the negative consequences of federal nonintervention in road building, the auto industry benefited tremendously from a lack of government scrutiny. The automobile business grew unfettered by state or federal regulation during the early years of the Progressive Era, even as new regulatory legislation crippled the giants of nineteenth-century industrialization. The first successful application of the Sherman Antitrust Act in 1904 broke up the Northern Securities Company, which, left untouched, would have created an unprecedented railroad monopoly. The Pure Food and Drug Act and the Meat Inspection Act, both passed by Congress on June 30, 1906, in the wake of the publication of Upton Sinclair's meatpacking exposé, *The Jungle*, established new food-safety guidelines and forbade the sale of many patent medicines, including a cocaine-infused concoction called Coca-Cola. Child-labor laws and compulsory education laws were passed first by states and then by the federal government with the Keating-Owens Act of 1916. Existing regulatory legislation was also strengthened. The Clayton Antitrust Act of 1914 replaced the weak Sherman Antitrust Act of 1890, and the Interstate Commerce Commission, which had been established in 1887 to regulate railroad shipping rates, finally gained adequate enforcement powers. Progressives also lowered the high protective tariffs passed under the Republican administrations of the 1890s, angering many manufacturers but pleasing farmers and consumers. American automakers dodged significant tariff decreases and watched their sales grow exponentially.[45]

Automobile executives worried, however, about unbridled competition. The number of auto manufacturers soared between 1900 and 1908, with some 485 new companies formed during that period alone. Barely half survived through 1908, but the frequency of entrances and exits in automaking highlighted the drawbacks of having neither internal industry standards nor external, government-mandated safety regulations. Greed motivated many inexperienced entrepreneurs to try auto manufacturing, and collectively they turned out a high volume of poor-quality, unsafe machines. Even among reputable automakers, the absence of safety standards made early model cars unreliable and limited in range.[46]

Against this backdrop of entrepreneurial chaos, the early auto industry relied on the leadership of a handful of innovative inventors and businessmen to improve the quality of its products. Michigan-born Henry Ford, the son of an Irish immigrant farmer, developed a talent for mechanical work as a young boy. In his thirties, after years of tinkering with internal combustion engines, Ford built one of the first gasoline-powered, self-propelled vehicles in the United States. With the help of investors, he became a pioneer automaker, the first to employ assembly-line mass-production techniques to build more-affordable cars. He began with the Model N in 1906, which sold for $600. Two years later, the more technologically advanced Model T changed the industry standard from heavy, elaborate touring cars to smaller and more practical machines.[47]

The Model T bridged the gap between the interests of urban and rural consumers. Because of its unique design and relatively low price, it was one of the first cars that farmers could use. The Model T sat high off the road, allowing it to clear rough terrain, and it used improved springs, making it both more durable and more comfortable over bumpy roads. As Ford refined his plant's production techniques, he lowered the price of the Model T from a starting price of $825 in 1908 to $680 in 1910, and then to just $345 by 1916. Thanks to Ford's ingenuity, cars became practical and affordable for city and country folk alike, and the company began marketing it nationwide as "the universal car." By 1918, half of the vehicles sold in the United States were Model Ts.[48]

Ford was not the only entrepreneur responsible for the car craze. Dozens of parts manufacturers reaped rewards by inventing vital new components and refining others to make cars safer and easier to drive. By

THE NEW FORD

THE UNIVERSAL CAR

The new Ford cars are up-to-the-minute in appearance, with large radiator and enclosed fan, hood with full streamline effect, crown fenders front and rear, black finish with nickel trimmings—a snappy looking car—and with all the dependable, enduring and economical qualities that have made the Ford "The Universal Car." One fact is worth more than a ton of guesses. Ford cars are selling from five to ten over any and all other cars, simply because they give more satisfactory service, last longer and are easier to operate and cost less to maintain—and there's no guessing about the reliability of Ford Service. Runabout, $345; Touring Car, $360; Coupelet, $505; Town Car, $595; Sedan, $645—f. o. b. Detroit. On sale at

Griffeth Implement Company
BROAD STREET.   ATHENS, GEORGIA

Ford Model T advertisement from the *Athens Herald*, November 25, 1916.

1901 more-comfortable steering wheels had replaced the tillers that were used on very early models, foot brakes worked better, and redesigned gearshifts made shifting smoother and more effortless. Open cars worked for touring elites who wanted to see and be seen, but dusty roads (and rainstorms) made them impractical for everyday use. In 1903 automakers began experimenting with closed cars, and removable canopy tops first appeared in 1904. Shortly thereafter, shock absorbers and bumpers became standard on many models. For added comfort, automakers lengthened wheelbases, installed better suspension, and moved seats slightly forward, and in 1907 they began using vanadium steel to make frames both lighter and stronger. The difficult and dangerous hand cranks used to start cars began to disappear after the first pneumatic self-starter appeared in 1907, and by 1912 modern electric self-starters replaced hand cranks on some cars. A number of improvements in the internal combustion engine also increased horsepower and fuel efficiency. And in the wake of the Model T, standard interchangeable parts helped other automakers compete with more reliable and affordable vehicles.[49]

While these advancements helped, tires still posed the biggest problem for rural driving. They were narrow and made of wood or hard solid rubber, designs that carved up dirt roads and were easily damaged. Wider pneumatic tires were first used in the early 1900s. Though expensive and not quite durable enough to withstand rough roadways, pneumatic tires provided drivers with a more cushioned ride and helped prevent further damage to the roads.[50] In 1905 Frank A. Seiberling, the owner of a carriage tire company in Akron, Ohio, introduced the first high-quality, easy-maintenance, straight-side pneumatic tire with a universal rim. His Goodyear Tire and Rubber Company, named after vulcanized rubber inventor Charles Goodyear, flourished when Henry Ford ordered 1,200 sets of Goodyear tires for the first series of Model T cars; by 1916 Goodyear was the leading tire manufacturer in the world.[51]

■ ■ ■

The spirit of innovation that fueled advances in automobile design would soon spill over into road-building technology and design. No individual embodied that trend more than Carl Graham Fisher. Fisher's humble origins did not portend his success as an automobile magnate and highway promoter, but his mechanical prowess and savvy salesmanship did. Raised in Indianapolis by a single mother, Fisher quit school in the sixth grade and worked a string of odd jobs to help support his mother and two brothers. By his teens, he had become infatuated with the bicycle craze and learned to ride the difficult, high-wheeled "ordinaries." The introduction of cheaper, mass-produced, low-wheeled "safeties" in the early 1890s made the sport more accessible to more people, so the enterprising Fisher opened a small repair shop when he was just sixteen. He wanted to expand into sales three years later but could not come up with the funds, so Fisher took a train to Toledo, Ohio, and talked top bicycle manufacturer Colonel Albert Pope into advancing him enough stock to open a store in Indianapolis. Pope, who would become one of Fisher's allies in the Good Roads Movement, reportedly told Fisher he had "more nerve than a government mule"—an assessment Fisher promptly corroborated by giving away fifty of Pope's bicycles in the first of many publicity stunts that made him famous in his hometown.[52]

As the bicycle fad gave way to the automobile craze at the turn of the century, Fisher transitioned along with it. For a few years, he made a good living racing cars in the Midwest, but after several deadly races and

Carl Graham Fisher, ca. 1915. (Courtesy of HistoryMiami, Miami, Florida)

mounting concerns about his deteriorating eyesight, he stopped racing cars and began selling them. Eager to outstrip his competition, Fisher staged outrageous publicity stunts, such as pushing a Stoddard-Dayton off the roof of a building to demonstrate its solid construction and sturdy suspension. The throng of onlookers either did not know or did not care that Fisher had ordered a one-of-a-kind reinforced car that could land in one piece, and he had let most of the air out of the car's tires to keep it from bouncing and sustaining damage. In his most famous stunt of all, Fisher attached an engineless car to a hot-air balloon, got behind the wheel, and floated over a cheering crowd in downtown Indianapolis. Among them was a pretty young teenager who was instantly smitten. Years later, Jane Fisher still remembered how invaluable the publicity had been, as well as the words her future husband had used to sell his product. "The Stoddard-Dayton was the first car to fly over Indianapolis," Fisher's ads read. "It should be your first car."[53]

Carl Fisher worked hard to persuade consumers that automobiles were safe and reliable, not just by orchestrating publicity events but also by manufacturing parts that transformed them into more-practical long-distance vehicles. Early-model automobiles did not have headlights, so all travel had to be completed during daylight hours. A driver who became lost or whose car broke down miles from home was stuck in the dark, and evening trips for work or leisure were impossible. Some motorists had tried hanging kerosene lanterns over the radiator, but they provided little illumination and, assuming the flames stayed lit, were dangerous as well. When a local inventor presented Fisher with an initial design for bright, reliable, compressed carbide gas-fueled headlights, he bought the rights to the patent and poured all of his time and money into fine-tuning them. With his business partner James Allison, Fisher's Prest-O-Lite company sold headlights as fast as his Indianapolis plant could make them and soon opened new plants on the East and West Coasts. The flammability of carbide gas caused many accidents and deadly explosions at the plants and resulted in numerous lawsuits against Fisher and Allison, but even the bad publicity could not hamper sales. In 1913, after less than a decade in business, Fisher and Allison sold the company and became millionaires several times over. Carl Fisher was not yet forty years old.[54]

Fisher invested much of his fortune in what became his most famous contribution to the auto industry: the Indianapolis Motor Speedway. Although the quality of automobiles had improved, American automakers still relied on their consumers to test the machines over time and distance. After years as both a participant and a spectator of automobile racing in the United States and abroad, Fisher had seen European automobiles outrun and outlast American cars time and again. Italian, German, and French cars, he complained, could "go uphill faster than the American cars can come down." On a trip to Europe in 1905, Fisher observed that European automakers tested their cars at high speeds over long distances on closed racetracks. American racetracks, by contrast, were far too short to permit cars to maintain high speeds for long periods of time, so American automakers did not have the same opportunity as European automakers to diagnose and fix mechanical problems. "I made up my mind then," Fisher recalled, "to build a speedway where cars could be run 1,000 miles in a test, if necessary." Four years later, he and three partners began construction on the Indianapolis Motor Speedway on a farm five miles outside of town. Automakers and drivers alike set up shop outside of Indianapolis in order to use the track as soon as it opened in 1909. Eddie Rickenbacker, the track's subsequent owner, later claimed that the speedway was responsible for most of the mechanical improvements to early cars.[55] Though probably an exaggeration, this was nonetheless a clear indication of how instrumental Carl Fisher was to the auto industry's efforts to build better and safer machines.

■ ■ ■

Fisher's interests shifted from building raceways to building roadways at precisely the moment that both auto sales and public support for the Good Roads Movement surged, prompting him and others to begin re-thinking old ideas about roads. He had been familiar with the Good Roads Movement since his bicycling days, but in 1912, while running for county commissioner in Marion County, Indiana, on the Progressive Party ticket, Fisher presented himself as a modern good-roads candidate. While cam-paigning among ordinary voters, he solicited support from auto-industry friends like Packard's Henry Joy for a transcontinental automobile high-way. His political opponents criticized the highway proposal, claiming that it would benefit only wealthy automobile owners, but Fisher insisted that automobile highways and farm-to-market roads were one and the same. He implored voters to spend more money for modern surfaced roads that would last longer. "I can not understand the folly of men," he complained, "who will throw thousands and hundreds of thousands of dollars annually into the gulf of oblivion, which the poor road really is." He asserted that "just a little money wisely expended goes farther and brings richer results than all the thousands that are thrown into the mudholes in the shape of loose gravel." Fisher promised to do his part by return-ing his $2,500 salary as commissioner and matching it with a personal contribution as long as the money went to the county road fund.[56]

Fisher lost the race, and the experience convinced him of the incom-patibility of good roads and local politics. Even during his campaign he had advocated "tak[ing] the road problem . . . out of politics" by central-izing control over roads at the state level, much like the administration of most school systems—an unusual argument, perhaps, from a candi-date running for local office. Writing to a friend shortly after the election, Fisher complained again, this time tongue-in-cheek, that "the highways of America are built chiefly of politics, whereas the proper material is crushed rock or concrete." The 1912 race would be his last foray into poli-tics, but not his last effort to transform the politics of local roadwork. For the next few years, Fisher used his reputation and his fortune to promote the construction of long-distance automobile highways and to urge state and federal control over roads.[57]

By the time Fisher began to imagine the Dixie Highway, auto sales had exploded among northerners and southerners and urban and rural consumers alike. Auto registrations nationwide increased by nearly 600 percent between 1905 and 1910.[58] Registrations in the South, which had lagged behind the rest of the country before the Model T, caught up by 1910 and continued to rise thereafter. In Georgia, automobile ownership rose more than 60 percent in a single year, increasing from 12,919 in 1913 to nearly 21,000 in 1914, and in North Carolina, ownership more than doubled that same year.[59] Masters of ingenuity, farmers put their new automobiles to work hauling crops to market and pulling plows, and they used the cars' engines to power cotton gins, corn shellers, and even generators to electrify their homes and barns. Cars also reduced rural isolation, even where poor roads limited the range of travel. By 1916 automobiles had become so indispensable to farmers that an agricultural official in Georgia declared that they had become "requisite[s] on the farm." Observers marveled at how quickly farmers had gone from criticizing cars to embracing them. "Once upon a time," the *Atlanta Constitution* opined, "the farmer and the automobile were . . . as incompatible as oil and water. But today farmers are among the foremost owners of cars."[60]

The automobile facilitated the uneasy alliance between farmers and auto men, but as both groups looked to the future, they knew that automobility would grow only as fast as highway mileage grew. As demand for cars expanded into rural areas and beyond, the scarcity of modern automobile-ready roads and highways replaced cost and quality as the auto industry's new challenge. But an even greater challenge would be figuring out how to build a seamless, integrated system of modern roads that served the needs of such a disparate set of consumers. On the eve of building the Dixie Highway, Carl Fisher's dream of bridging North and South, urban and rural, seemed as impossible as it was visionary.

Farmers, Auto Men, and the Highway Lobby

Visions of modern highways became clouded when politics entered the mix. Joining old local roads into a network of paved highways meant balancing not only the practical transportation needs but also the political interests of auto manufacturers, local governments, and rural citizens. Each of these groups had its own set of expectations, and each harbored its own unique understanding of the role the government would play in the process. Auto men wanted federal money but not federal regulation of their business practices. State and local officials welcomed federal assistance yet cautiously guarded their own control over regional affairs. And although farmers eagerly embraced federal funding for new roads, they perhaps more than anyone understood what a great—and expensive—challenge it would be to link their isolated communities with modern automobile highways.

At the most basic level, all of these groups agreed that modern roads required federal funding, particularly the auto industry. Speaking before the first meeting of the American Road Congress in late 1911, Chalmers Motor Company president Hugh Chalmers declared: "What the automobile industry most wants to see is the federal government actively enlisted in the work of improving our roads." He opined that "there is no one single thing that a State or nation can do" that has "such a sure and impartial benefit to all the people as to build good roads." Chalmers argued that the monumental job stood outside the capability of even wealthy and powerful automakers. "The most that the automobile industry can do in the way of furthering good roads is . . . carry on missionary work—try to mold public sentiment so that it will take action." It would take "big scale" federal support to build modern highways. Chalmers called for uniform national laws governing everything from licensing automobiles to speed limits and even advocated specific new tax legislation to help pay for modern long-distance, automobile-ready roads and highways. And he believed that "when the people . . . realize that the benefits are not for one class of people alone, but for all the people alike, that they will rise up some day and demand of the national congress . . . and the State assemblies . . . that they cooperate" to build better roads.[61]

After 1910, auto-industry pioneers capitalized on the opportunity to shape federal intervention in the nation's fastest-growing industry through participation in the Good Roads Movement. After a decade of working to make automobiles into long-distance vehicles, the automobile industry emerged as a vocal and visible leader in the good-roads campaign. They wrested control of the conversation away from the federal Office of Public Road Inquiry, renamed the Office of Public Roads (OPR) in 1905, and its railroad sponsors and increasingly blurred the distinction between city streets and country roads as urban traffic bled into the countryside and farmers ventured into towns and cities. Their message was that automobiles promised both urban tourists and farmers the freedom to move over longer distances more cheaply and efficiently than railroads ever did. They promoted the construction of long-distance, or "trunk line," roads that supplanted, rather than supplemented, railroads.[62]

Still deeply committed to the railroads, the federal government was slower to embrace this new vision of road-based transportation. Congress steadily increased the OPR's funding, from $50,000 in 1905 to $279,000 in 1914, yet it did not grant the agency any road-building powers.[63] During that same period, congressmen warmed to the idea of federal highway aid, but they made no coordinated effort to determine which roads to fund, how much aid to give, or where the money would come from. Instead, they proposed a mishmash of highway bills, each one with a different plan and a different price tag. Southerners proposed many of these bills, including the popular Brownlow-Latimer Bill, a $24 million proposal to funnel federal aid through the states. Although the bill had wide-ranging support, it never made it to a vote.[64] Mississippi representative Ezekiel S. Candler introduced two bills, including a massive $100 million proposal to build and improve rural postal routes, but he skimped on the details, arguing that they could be worked out later.[65] In 1912 the authors of several of these bills cooperated on what became known as the Shackleford Bill, a complicated $25 million proposal that would grant federal aid for rural post roads but preserve local and state control over them. The bill passed the House two years later with near-unanimous southern support (and much northern urban opposition), but it fizzled out in the Senate.[66] No fewer than sixty road bills of all different varieties stalled in Congress in the seven-month period between December 1911 and July 1912 alone. Except for a modest $500,000 increase in the Post Office Appropriation Bill of 1913 for post roads, Congress did not commit any funding to road construction.[67] As such, most representatives lagged far behind popular interest and support for federal funding for new roads and highways. The Good Roads Movement's campaign for more roads and highways intensified as the automobile steadily upped the ante in the campaign for more and better roads. Good-roads organizations nationwide numbered more than 450 by 1912, forming a powerful private-sector lobby.[68]

■ ■ ■

At the local level, enthusiasm for better roads—and the government aid they required—had been building for years, particularly in the South. A major turning point was the advent of short-demonstration, or "object-lesson," roads. Since ORI chief Martin Dodge first proposed them in 1900, object-lesson road projects brought federal engineers in to teach county officials newer and more-scientific road-building techniques. When engineer Logan Waller Page took over the OPR in 1905, he took advantage of the agency's increased budget as well as the momentum of the Good Roads Movement to revive object-lesson roads.[69]

Because counties had to pay the actual construction costs of object-lesson roads, responses to the OPR's offer varied according to local resources. In 1902 Fulton County, Georgia, home to the state's largest city, Atlanta, was already building its own model roads with the help of good-roads organizations and railroad companies. Describing a recent good-roads convention that included an object lesson, a Fulton commissioner boasted to federal officials that the county had been building its own "most excellent roads . . . both before and since" it began building instructional roads. But he added that "this is far from true in the rural Counties, where the population is less dense" and whose road officials benefited from seeing the object lessons in their larger sister county.[70] When the OPR offered the county a *federal* object lesson three years later, the chairman of the county's Public Works Committee rebuffed it. In a curt reply, he declared that "the roads in this County are built under strictly scientific principles, and there is little need of availing ourselves of this offer at this time."[71] Whether the county viewed the OPR's offer as an infringement on local control or just as an unnecessary expense, the response indicated that urban officials did not see any critical need for federal aid for local roads.

Object-lesson roads proved more appealing to rural southerners, who were deeply dissatisfied with the limitations of local roads and local control, but they were in no position to pay for them. Wayne County, Georgia, had no improved roads, so county commissioners were eager to take advantage of the OPR's offer. However, they also had no road machinery and only the most primitive tools and materials to work with. After seeing an advertisement about object-lesson roads in *Southern Agriculturalist*, commissioner J. F. Surrency asked the OPR for details about the labor, materials, and number of teams needed because he could not assume the county would be able to provide the basic materials required by the OPR's expert engineers.[72] A city official in the tiny town of Fitzgerald also contacted the secretary of agriculture for help after seeing the same advertisement. "It is a question," he declared, "if any place in the United States is in greater need of roads than south Georgia." But he, too, wondered about the cost.[73] E. L. Bardwell of Talbot County might have argued that middle Georgia was no better off. He contacted the OPR to ask for general advice about building and maintaining roads "in a sparsely settled, agricultural county . . . [with] no town of more than 12 or 1,500 population . . . [and whose] bad roads are sometimes rendered nearly impassable by the winter rains." Bardwell's detailed list of questions about the cost of roads and the personnel needed to build them implied that there was no regular system of road construction or maintenance in the county.[74] Dozens of other rural officials and ordinary citizens wrote to the OPR asking for help, only to discover that they could not afford to actually apply for an object-lesson road. To the great frustration of OPR officials, most people failed to follow up after an initial inquiry about federal assistance. For example, a Dougherty County commissioner in the south Georgia town of Albany wrote to the OPR in December 1908 requesting expert assistance in building expensive macadam roads "at earliest moment." The OPR sent the commissioner an application, but five months later, he still had not returned it. In May the office sent another letter urging him to apply "at once" if he hoped to have an engineer sent down before the end of the year, but he never did.[75] However popular an idea it was, Page's revival of the object-lesson road program generated very few actual road improvements. By 1912 the OPR had supervised the construction of only 616 miles of object-lesson roads nationwide, and 500 of them were made of dirt.[76]

Still, the object-lesson road program was a critical stepping-stone in the Good Roads Movement because it highlighted for Congress the OPR's impotence as a partner in the campaign for federal highway aid. Citizens often appealed directly to senators and congressmen to ask about rumors of federal help for local roads. These inquiries gave the OPR greater leverage in its ongoing struggle with Congress by forcing legislators to take some responsibility for their refusal to fund federal roadwork. When U.S. representative Thomas Bell wrote to the OPR on behalf of his constituents in Ellijay, Georgia, asking that the agency provide not only an engineer but also a rock crusher and a roller to build a hard-surfaced road, an OPR official seized the opportunity to remind Bell of the agency's limitations. "Congress does not authorize us to purchase or rent road building machinery," he pointedly reminded the congressman, "and, in consequence, we have none to furnish to assist in the work at Ellijay, and therefore cannot grant the assistance which is desired."[77]

This arrangement also underscored the influence of local citizens, who understood all too well just how political road building was. They may not have been able to get the help they needed from the OPR, but they could exert pressure on their representatives in Washington. Joel Deese of tiny

Cochran, Georgia, wrote to Congressman Dudley M. Hughes complaining about local road officials. The year before, Deese said, he and some of his neighbors had campaigned against "expensive and inefficient conduct [of] road construction" by county officials, and they succeeded in replacing local road officials with men committed to newer, better methods of construction. But the new commission was not up to their standards, either. Deese asked Hughes to publish letters in the local newspapers telling readers about the availability of federal aid through the USDA and the OPR, a direct appeal to county residents that Deese hoped would "spur our officials to take notice of an active public demand" for more-aggressive roadwork. He implored Hughes to intervene further by sending a federal engineer to investigate their work, and he gently reminded the congressman that his reelection prospects depended on his response. Hughes's predecessor, Deese said, had refused, "fearing, doubtless, that it would queer his campaign; which it undoubtedly did, though not in the way he feared." In a long and testy reply, Hughes defended his record as a good-roads candidate. He complained that he had implored all of the counties in his district to take advantage of the OPR's offer, but only one did. Declaring that he had done all he could do, Hughes patently refused to intervene in local affairs. "This should be a question that the county alone should adjust," he instructed Deese. "I think outside interference would be justly criticized."[78]

Like so many other congressmen, Hughes did not understand the crux of the disagreement between local residents and their road commissioners. It was precisely because local officials could not afford the minimal aid offered by the OPR that voters like Deese felt obliged to go over their heads. By writing to their congressmen rather than to the OPR, rural road advocates forced them to become intermediaries in the relationship between local and federal road officials. Though detached from the debates between OPR officials and Congress, ordinary citizens fully comprehended the association between OPR assistance and the elected officials who funded it. It may not have induced federal representatives to intervene in road construction, but it showed clearly that local people wanted them to. And even if they did not grant the OPR more authority and sufficient funding, congressmen learned more about the accomplishments and the limitations of the subagency through this correspondence. Moreover, bringing elected federal officials into OPR business politicized the office in ways that Congress and the Department of Agriculture had explicitly forbidden. This tested the boundaries of the OPR's authority and politicized the road issue at the federal level, something the very creation of the office had been engineered to avoid.

Harnessing public enthusiasm for more-aggressive federal intervention, manufacturers, auto enthusiasts, and road experts began to form more-powerful lobbying groups. Auto-industry insiders established a number of organizations dedicated to direct lobbying for federal funding for highway construction. The industry's first trade association, the National Association of Automobile Manufacturers (NAAM), was formed in 1900 with the specific intent of organizing and coordinating the industry's attempts to influence favorable legislation. So, too, were the American Motor Car Manufacturing Association (AMCMA) and the Association of Licensed Automobile Manufacturers (ALAM). The most influential of these organizations was the American Automobile Association (AAA), which forged strong ties with the southern Good Roads Movement.[79]

Expert road builders and frustrated local road officials formed their own groups as well. Among them were the American Road Makers (ARM), later called the American Road Builders Association, whose diverse membership included contractors, engineers, manufacturers of and dealers in road machinery, county commissioners, and a handful of state highway officials.[80] Even OPR chief Logan Page, whose efforts to remain impartial and apolitical were short-lived, presided over the American Association for Highway Improvement (AAHI), an umbrella organization for the Good Roads Movement that lobbied for state and federal highway aid. Along with other major organizations such as the AAA, the AAHI sponsored annual meetings of the American Road Congress, where road officials and auto industrialists met to debate the merits of proposed state and federal legislation. Page even escorted AAHI members to congressional hearings so they could lobby congressmen face-to-face for favorable road legislation.[81]

With new lobbying power and a Congress in favor of some form of federal aid, highway supporters turned to the media to help proselytize the gospel of good roads to the masses. Auto industrialists sponsored long-distance tours called "reliability runs" to publicly promote a new model of highway-based transportation for farmers and automobile tourists alike. With the help of wealthy benefactors such as Boston telephone pioneer Charles J. Glidden, they staged city-to-city tours designed to simultaneously demonstrate the practicality of interstate routes and market the automobile as a dependable means of long-distance transportation. The Glidden Tours, which ran from 1905 until 1913 and were organized by the AAA, were front-page news in local and national newspapers, covered by dozens of reporters who tagged along on the arduous journeys and recorded every detail. Their accounts described encounters with people who had never seen a car before, welcoming committees in the countless small towns they passed through, and endless frustration with vehicles that frequently broke down along rough terrain or became mired in thick, glue-like mud up to their axles.[82]

The AAA organized the first Glidden Tour through the South in 1911, and it proved just how adept the industry had become at simultaneously marketing automobiles and lobbying politicians. Eighty-six cars left New York City on October 14 bound for Jacksonville, Florida, and led by Georgia governor Hoke Smith, the first governor ever to participate in the Glidden Tours. Smith's participation was a clear indication that his views on road building were not incompatible with those of the AAA.[83] The drivers had few major problems until they reached Virginia, where, chirped one reporter, "Glidden tourists . . . gave good-bye to good roads." Teams of mules had to wrench numerous cars from the mud. Hoke Smith's car skidded off a bad road in the rain and hit a telegraph pole.[84] Reporters were merciless in their descriptions of poor road conditions in the South, with one declaring: "[N]othing has approached in vileness the condition of the roads in Virginia." Still, that same reporter concluded that the tour was a showcase for the durability of modern automobiles. "Two years ago all except a comparatively few cars would have refused to even attempt" the treacherous trip over southern roads, he said. "It was astonishing," he continued, "to see small motor vehicles plowing through mud three feet deep on a steep up-grade, and touring cars with full complements of passengers [doing] all this and more."[85] Following the tour, at an AAA dinner attended by several governors and legislators, an AAA official optimistically predicted that the South would soon "outstrip" the rest of the country building roads.[86]

Numerous other tours were sponsored by newspapers and businessmen eager to capitalize on the popularity of the Glidden Tours. In 1909 the editors of the *New York Herald* and the *Atlanta Journal* sponsored a trailblazing tour to select a route for the Broadway-Whitehall National Highway between their two cities. Reporters accompanied the caravan and telegraphed back every detail, from the gleaming modern streets of New York City to the pitted, often impassable rural roads below the Mason-Dixon Line. Their accounts of "auto tourists in peril" were as exciting as any dime-store novel. Famous participants added to the appeal. The first car to reach Roanoke on the tour carried baseball star Ty Cobb, who drew a huge crowd.[87] Readers could not get enough. *Atlanta Journal* editor Major John S. Cohen declared the proposed highway an opportunity for "the elimination of the final barrier" between North and South. The tour generated good publicity and sold a lot of papers, but when it was finished, so was Cohen. "Now it remains for the counties themselves to finish the work," he concluded, "to build up the highway, to make it what it has just begun to be."[88]

The newspaper stunt did not generate a modern highway, but it did attract the interest of other enterprising capitalists. Upon hearing that the newspapers would be sponsoring an automobile tour, a North Carolina millionaire named Leonard Tufts waged a spirited campaign to persuade the editors to select a route through Washington, D.C., through the state capitals of Virginia and the Carolinas, and on to Atlanta. Conveniently for Tufts, his "Capital Highway" passed through the village of Pinehurst, North Carolina, where he owned a tourist resort that his wealthy northern customers could reach only by railroad. Tufts did not get his highway, but his proposal generated invaluable publicity for both his business and his region's highway needs.[89] The following year, Georgia businessmen sponsored a "Round the State Tour." Like earlier tours, this one did not generate new roads or new funding, but it did strengthen support for new long-distance highways in the South.[90]

Yet the widespread interest in auto highways put some southerners on guard. Wary that roads would privilege the leisure interests of the wealthy over the practical demands of rural farmers, southern politicians began to articulate concerns over the direction of the highway enterprise. Perhaps no one articulated such reservations better than Hoke Smith. Like Henry Grady, Atlanta's most prominent New South evangelist, Smith was a newspaper editor and an ardent supporter of southern economic development. However, Smith built his political career on helping farmers, not industrialists. As both an attorney and an elected official, Smith was famous for fighting the railroads. He owed the rest of his political career to the promotion of rural white political interests as well. Smith won the race for governor of Georgia in 1906 by promising to disfranchise African Americans, a popular "electoral reform" in much of the Progressive

Era South. As a senator, he sponsored the Smith-Lever Act of 1914, which established the Agricultural Extension Service, and cosponsored with Georgia representative Dudley M. Hughes the Smith-Hughes Act, which funded vocational programs and was an important source of federal aid for rural secondary schools.[91] Although he had tipped his hat to the AAA's goals for the Good Roads Movement by participating in the 1911 Glidden Tour, Smith remained circumspect about the priorities of the good-roads campaign.

At a 1914 meeting of the American Road Congress, Smith made it clear that there remained a yawning gap between the interests of northern and southern good-roads progressives. Smith supported federal aid for roads, but he persisted in differentiating between rural market roads and automobile highways. In his address, he expressed his fear that "charming advocates of national highways" might make supporters of the Good Roads Movement forget the importance of also building roads for farmers. Smith declared his preference for funding rural market roads over major tourist highways. "Nobody enjoys a great national highway or a Glidden tour more than I do," he said, "but their chief value is that they may stimulate better roads away from the great highways."[92]

Quick to point out the limitations of this thinking, AAA president George Diehl argued that like so many others who still made distinctions between farm roads and "peacock alleys," Senator Smith had erred in assuming that the AAA promoted impractical routes for only auto tourists. Instead, Diehl argued that the federal government should fund the construction of "important" highways first. He hoped, not unlike Smith, that these "great national roads and State highways will form the backbone from which will radiate the county roads and the township roads and the whole will be combined into one properly connected and well developed system of highways, which together with the use of motor vehicles, will make for the greatest development this country has ever known." Although he denied that the association had any specific interest in "ocean to ocean" or "Great Lakes to the Gulf" highways, Diehl had outlined a vision of modern highways that rested on long-distance highways that anchored the nation's transportation system and served as both automobile routes and major market highways.[93]

But the divisions were there and would remain a source of friction for some time. Although both Smith and Diehl fundamentally believed in the importance of roads, regional politics and deep-seated ideological differences about the role of government in the lives of citizens began to sow the seeds of conflict. Neither would have argued that modern highways should be built. But dogged questions—Who would these roads benefit? Who were they for? Who would control their construction? What purpose would they serve? Where would they be built? And who would reap the political benefits when all was said and done?—began to muddy the clarity and unity of their purpose. As enthusiasm for large-scale modern highway projects began to take shape, so too did the roots of political discord.

Nevertheless, by 1914 the unlikely alliances that underpinned the Good Roads Movement had forever changed the way people thought about roads. After years of disappointing efforts to build roads at the local level, ordinary citizens, auto businessmen, and public officials had banded together to rethink the scope and purpose of individual transportation. With automobiles here to stay, good-roads advocates forged ahead with plans that would leave an indelible mark on the nation's highway system and its system of government.

The Road to Dixie, 1914–1916

In November 1914 Carl Fisher's proposal to build the first modern automobile highway linking the North and the South provoked a jubilant response. Southern newspapers rushed to endorse the so-called Cotton Belt Route, declaring it "a splendid investment" that would encourage the construction of more roads. Northerners rejoiced at the news as well. Northern tourists, said one Indianapolis man, "have longings in their hearts to tour the great south . . . to motor into its lands of tradition and history."[1] The new highway would be "of great and mutual benefit to the people north and south," according to Governor Samuel Ralston of Indiana, "and an ever strengthening bond of unity between the people of these states." The *New York Times* dubbed it "The Dixie Peaceway" and declared the route "a memorial . . . symbolical of the accord between brethren which shall never again be broken."[2] The highway itself seemed poised to remedy the conflicts of the past and transport the nation into a new era of sectional harmony.

Soon renamed in honor of the region it promised to transform, the Dixie Highway became the most successful and consequential of the so-called marked trails of the 1910s and 1920s. These privately planned routes used signage to link existing local roads into the first long-distance highways in the nation. The meandering, unpaved routes were difficult to navigate, but in an age when railroads still dominated interstate travel, these primitive freeways captured the nation's imagination. They became symbols of modern-day progress that would heal the lingering wounds of the Civil War. Yet some marked trails evoked that past even as they heralded the future. The transcontinental Lincoln Highway between New York and San Francisco honored the fallen president who had preserved the Union, while several regional highways in the South, such as the Jefferson Davis Highway in Mississippi and the Robert E. Lee Highway in Virginia, paid tribute to the old Confederacy.[3] The north-south Dixie Highway, however, merged the two in a quintessentially modern monument to national unity. Most marked trails were either untenable from the start or, in some cases, deceitful schemes to extract donations from gullible citizens along the route. Few were ever finished. But Fisher's visionary route, far longer and more complicated than any other highway of its time, was completed within a decade.

The Dixie Highway galvanized support unlike any other marked trail in part because it embodied the spirit of cooperation and unity that the nation so badly craved in the new century. Supporters believed that the Dixie Highway would bring people together because it benefited everyone. Northerners and southerners, city people and farmers, and governors and county commissioners all rallied behind the campaign to build the highway. But as they would soon discover, working together was not always easy. Merging the priorities of so many different groups proved so challenging that it nearly derailed the Dixie Highway. Intense competition over routing decisions divided local communities even as it generated publicity and funding for the highway. Self-interested businessmen tried to use their powerful political connections to influence the route, while governors of the Dixie Highway states were powerless to stop them. County road authorities promised to devote resources they did not have to begin building a highway that they hoped state and federal governments would someday finish. Supporters established a formal organization to guide the project through the routing difficulties, but the officers of the Dixie Highway Association (DHA) struggled to reconcile competing interests after the final route was selected. Working together in the spirit of cooperation proved more divisive than anyone could have imagined.

Within a year of the Dixie Highway proposal, it became clear that envisioning a modern interstate highway was simple, but building it one county at a time was nearly impossible. Enthusiasm for the highway never waned, but no single, clear agenda emerged from the motley mix of Dixie Highway communities. The paeans to sectional reconciliation that had marked the announcement in 1914 soon gave way to pitched battles that exposed the very real political and economic differences that still divided North and South and rural and urban. In response to these problems, the DHA took up the Good Roads Movement's crusade for federal highway aid. They believed that the regional, economic, and political interests of so many different parties could only be unified with the financial and bureaucratic intervention of the federal government. As they worked to help counties complete their portions of the Dixie Highway, DHA officers launched a parallel campaign to win federal funding for vital interstate routes like theirs. Thus even while local tensions over routing, regional differences in funding, and clashes between the business and the politics of road building shaped the Dixie Highway, the Dixie Highway shaped the future of modern highway construction.

Uniting North and South

Early marked trails like the Broadway-Whitehall Highway between New York City and Atlanta were simple unpaved routes linking two major cities, difficult to travel even in good weather and rarely connected to other long-distance roads.[4] Few motorists used them, and those who did regretted it. As automobile sales spread beyond urban areas and tourists ventured deeper into the countryside, these primitive thruways proved increasingly unsuitable. In September 1913 the South's leading business weekly, the *Manufacturers Record*, proposed a bold alternative to the marked trails. In place of scattered and disconnected dirt roads, the paper suggested a "comprehensive" system of interconnected routes linking all of the regions in the country. "The highway demanded by the times," the paper declared, "is one unbroken line from the North and the West through the South, built so solidly that it cannot be affected by weather."[5]

This visionary suggestion for an integrated system of modern roads has been called the inspiration for the Dixie Highway, but more accurately it was a prescient blend of two very different highways first proposed by Carl Fisher. The only one already under way in 1913 was the Lincoln Highway, a transcontinental route from New York to San Francisco named in honor of Fisher's childhood hero.[6] Fisher and several other auto industrialists believed they could build the highway quicker and more efficiently by using private donations. But private capital could not forestall the influence of local politics. During a month-long pathfinding tour of the West in 1913, nicknamed the "Hoosier Tour" by the press, Fisher and other members of the Lincoln Highway Association (LHA) quarreled over routing options. Their disagreements fueled rumors and public speculation about where the highway would go. Ever the salesman, Fisher exploited these rumors in hopes of raising more donations from people along rival routes. His plan backfired when the LHA made the official routing announcement on August 26, 1913, and thousands of supporters and donors learned that the Lincoln Highway would bypass their communities altogether. Some badgered the LHA to add their towns to the route, while others simply refused to support the highway any longer. Controversy and disappointment would hamstring the LHA's efforts to complete the highway for years to come.[7]

Harnessing private capital to build public highways turned out to be impractical. Despite great enthusiasm for the Lincoln Highway, donations fell far short of the $10 million Fisher had solicited. Months after the official route was selected, the LHA had pledges for barely half that amount and even less in hand, and most of the rough dirt roads that formed the Lincoln Highway west of the Mississippi River remained untouched. When a small-town newspaper editor from Nevada wrote to the LHA and demanded to know what the organization had done with the donation money, he voiced the doubts and frustrations of thousands of disappointed westerners. Perhaps if the man had known that the LHA had raised only $265 in all of Nevada he would have understood why improvements along the highway had stalled. Jokes about the Lincoln Highway soon replaced the early praise. Some called it a line on a map connecting the nation's worst mud holes. In 1915 writer Emily Post famously reported that when she asked what was the best road to California, a friend replied without hesitation, "the Union Pacific." The Lincoln Highway, her friend declared, was just "an imaginary line like the Equator!" Even Carl Fisher gave up on the highway. After 1915 he did not play an active role in the LHA.[8]

The Lincoln Highway was, in a sense, a trial run for the Dixie Highway. Despite the project's shortcomings, the sheer ambition of the Lincoln Highway inspired both capitalists and ordinary citizens to think differently about the future of transportation after 1913. The Baltimore-based *Manufacturers Record* saw the flurry of publicity surrounding the Lincoln Highway as an opportunity to prod readers into demanding a similar highway through the South. A north-south highway, the paper insisted, was "equally as important—indeed, we believe far more important" but required "the energetic co-operation of the business interests of the North and West in connection with those in the South." The paper urged "the great capitalists of the South" to follow the lead of the auto men in the

Lincoln Highway Association and mount a southern highway campaign. The readers who flooded the *Record*'s office with letters of support could not have known that Carl Fisher would soon take on that responsibility by bringing together automobile businessmen and regular citizens in a campaign to build the Dixie Highway, drawing on the hard lessons he learned from the Lincoln Highway experiment.[9]

Carl Fisher's real estate ventures in south Florida prompted his interest in building a north-south highway. He had first visited Miami during his honeymoon in 1909 and was so taken with the sleepy beach town that he purchased a vacation home there a few months later, sight unseen. From her new home, Jane Fisher could see a mosquito-ridden "strip of jungle" just across Biscayne Bay, accessible only by boat. Carl, however, envisioned a city there. He bought up hundreds of acres of land and promptly invested $50,000 to build a two-and-a-half mile wooden bridge from the mainland. But building a city proved more difficult. "Crazy Carl Fisher," people said as the work dragged on and expenses soared year after year, was "sinking his millions in mud." Undaunted, Fisher cheerfully gambled both his personal fortune and his reputation to finish what he had started. Over the next several years, teams of mules, men with machetes, and fleets of dredging ships transformed a dense mangrove swamp into a sparkling resort city that Fisher named Miami Beach.[10]

By 1914, a full decade before the Florida land boom transformed the rest of the state, Fisher was ready to market his beachfront properties to wealthy tourists. He courted some of the giants in the auto industry, including Billy Durant of General Motors, Edsel Ford, and executives at Packard, Willys-Overland, and Firestone.[11] In his typical extravagant fashion, Fisher reasoned that there was no better way to sell Miami Beach to tourists than to build a tourist highway right to Miami Beach. Other proposals for a north-south route, including that of the *Manufacturers Record* a year earlier, had stalled, but Fisher had enough clout with wealthy and influential auto industrialists to launch a high-profile campaign. He floated the idea to several friends in the Midwest, who agreed that the automaking region was the ideal starting point for an automobile route. However, his experience with the Lincoln Highway had taught Fisher the limits of the auto industry's leadership, so he also called on his friend Governor Ralston of Indiana, to lend his name to the new highway proposal. Further aware that his personal stake in a highway between the Midwest and Miami Beach might be injurious to the project, Fisher shrewdly stepped aside and tapped an unknown disciple of the Good Roads Movement to be the spokesman for his latest business venture.[12]

No one had heard of Fisher's pick—Indianapolis seed manufacturer William S. Gilbreath—when he arrived in Atlanta for the fourth annual meeting of the American Road Congress in early November 1914, but his name would be forever linked to the Dixie Highway. As secretary of the Hoosier Motor Club and an ardent disciple of the Good Roads Movement, Gilbreath knew the meeting would be an ideal backdrop for his announcement about the new highway project, particularly since it was being held for the first time in the Deep South, where good roads were most scarce. Through the Atlanta newspapers, Gilbreath introduced plans for the "Cotton Belt Route," which would link major cities between Chicago and Miami, sending hundreds of thousands of northern tourists through the South and generating much-needed revenue for the region. A letter of introduction from Governor Ralston helped Gilbreath secure a meeting with Georgia governor John M. Slaton, who pledged his cooperation, and Gilbreath vowed to get endorsements from the other governors whose states the highway would traverse.[13] The announcement received little attention outside of the South, but then Carl Fisher had learned his lesson about promoting highways. While the Lincoln Highway still dominated national headlines, Fisher and Gilbreath crafted a careful, methodical, and very different plan for the Dixie Highway.[14]

The Dixie Highway diverged from other marked trails in several important ways. The Lincoln Highway had faltered under the unilateral control of auto men, but the Dixie Highway incorporated state leaders from the very start. A month after the highway announcement, Governor Ralston of Indiana invited the governors of the other Dixie Highway states to meet halfway in Chattanooga the following April in order to establish a highway association to oversee the project. Fisher, Gilbreath, and the Hoosier Motor Club boosters who coordinated initial plans for the Dixie Highway did not solicit donations, speculate about the route, or even formally adopt the name "Dixie Highway." Instead, they left all major decisions up to the governors.[15] They nevertheless moved quickly to organize the project. Whereas plans for the Lincoln Highway lingered for years and other marked trails never made it past the proposal stage, the Dixie Highway evolved into a viable route within a few short months.

One reason for this was the expectation of government aid for the highway. By 1915 arguments against federal intervention in road building were

scarce, even in the most rural sections of the South. A Dixie Highway supporter in the south Georgia town of Fitzgerald believed that it was only a
matter of time before the federal government exercised its "constitutional
right, placed there by the founders of this Government, of constructing
a system of permanent highways." Pretty soon, he predicted, millions
in federal appropriations "will . . . be diverted to the inter-state roads
of the country," and "these highways will become the main arteries of
travel."[16] Fisher cultivated this impression as carefully as he orchestrated
Gilbreath's role in the announcement. All of Gilbreath's early public statements implied that Indiana governor Ralston helped conceive the idea
of a north-south highway, and that he, Governor Slaton, and the other
governors would see it through. Timing the announcement to coincide
with the American Road Congress also permitted Fisher and Gilbreath to
capitalize on discussions about state and federal highway aid that were
taking place during the conference. Many of the road experts who spoke
at the meeting in Atlanta promoted federal aid as well as the expansion
of state aid to states that did not yet have it. The entire afternoon session on November 9 was devoted to federal aid, with U.S. congressmen,
state officials, representatives of the American Automobile Association,
and officers of regional good-roads associations outlining their reasons
for supporting national highway legislation. Dr. Joseph Hyde Pratt, state
geologist from North Carolina, confidently predicted that it was "simply a
question of a short time" before Congress would pass a federal aid bill.[17]
Gilbreath's Atlanta audience must have shared Pratt's optimism when
they read the next day's paper, which featured Gilbreath's front-page
announcement for the Dixie Highway alongside a headline declaring:
"[Road] Congress to Ask Government Aid for Good Roads." Over the next
several weeks, publicity for the highway reinforced these associations
over and over again.[18] By casting the Dixie Highway as a government
road just when support for state and federal funding was cresting, Fisher
and Gilbreath guaranteed it a degree of legitimacy missing in most other
marked trails.

Another reason the Dixie Highway proposal gained momentum so
quickly was its economic appeal. Marked trails naturally attracted automakers and the tourist industry, but the Dixie Highway drew interest from
a far broader group of regional and local business interests. Unlike the
Lincoln Highway, which traversed hundreds of miles of sparsely populated areas in the West, the Dixie Highway linked heavily populated areas

of the Midwest and South. Even before the exact route became clear, anyone could see that it would link several major cities with dozens of sizable market towns. The highway would generate immediate business for a wide range of people, from restaurant and hotel owners in larger towns and cities to merchants and filling station owners in small towns and farmers operating roadside produce stands in the countryside or hauling crops to market. Local business clubs and chambers of commerce wasted no time lobbying the governors for spots on the new route. Scarcely a month after the Dixie Highway was announced, a representative of the Charleston, South Carolina, chamber of commerce urged the governor of Georgia to consider supporting a detour through the Lowcountry.[19] The president of the Monteagle Board of Trade petitioned Tennessee governor Tom C. Rye, arguing that a highway through Monteagle would "add thousands per cent to the already wonderful popularity of this section as a summer resort" as well as promote the farming interests of the nearby Cumberland Plateau.[20] More-prominent businessmen, like Charles E. "Charlie" James, a Chattanooga real estate developer and the owner of the Signal Mountain Inn, were also among the earliest and most active boosters of the Dixie Highway.[21] Even small-town businessmen recognized the highway's potential. A. A. Womack of the Manchester Commercial Club wrote to Governor Rye to promote a route through his town, which he said offered "the finest scenery between Nashville and Chattanooga, the best farming lands in Middle Tennessee," and "excellent accommodations for tourists."[22]

Yet the campaign to promote the highway grew so large so quickly that it obscured some significant flaws in the original proposal. Carl Fisher's carefully staged announcement had not only concealed the true origins of the highway idea, but it also had exaggerated the likelihood of immediate state or federal funding for it. Despite ongoing efforts to pass comprehensive federal highway legislation, the American Automobile Association, highway engineering experts, and politicians could not find common ground on a single bill. Led by the AAA, the highway lobby continued to press for a national highway system, but bold and expensive proposals such as Washington congressman Stanton Warburton's bill to link all of the nation's state capitals, large cities, and seaports with a system of paved interstate highways went nowhere. Neither did more-conservative proposals that would have had little or no practical effect on roads. In 1914 no fewer than forty-nine bills were introduced in Congress, many of them, according to one critic, "intended rather for political effect than as measures intended for enactment into law."[23] There was little reason to think the federal Office of Public Roads would give even token support to the Dixie Highway, either, because OPR officials were suspicious of marked trails. In 1912, for instance, they had denounced the Battlefield Route between Chattanooga and Atlanta as a "scheme," though it would later become one of the most promising potential routes for the Dixie Highway.[24] Despite the roles that governors played in the project, state aid was also no guarantee. At the time of the Road Congress, only two states between Illinois and Florida had state highway commissions: Illinois and Kentucky. There were none south of Kentucky, and even Kentucky's highway commission served only in an advisory capacity. Gubernatorial endorsements from states that had neither highway commissions nor laws permitting state expenditures for roads were nominal at best.[25] And allowing the governors to select a route for the Dixie Highway did not change the fact that the highway was really a string of local roads. Privately, even the governors themselves remained equivocal in their support for the highway. Both Governor Rye of Tennessee and Governor Slaton of Georgia doubted they would attend the governors conference at all.[26]

Most people supported Fisher's proposal for a north-south highway because they wanted to believe in it, not because they were duped into thinking it would be easy to build. No one had a blueprint for modern highway construction. During the Road Congress in Atlanta, speakers from the Dixie Highway states expressed support for long-distance highways but contradicted one another over how best to build them. In his welcoming address, delivered just hours after he endorsed the Dixie Highway, Governor Slaton praised the potential of modern highways to "annihilate" distances and unify people but admitted his own powerlessness to build them. Instead, he told the assembled road experts to direct their expert advice to the county commissioners still in charge of the state's roads.[27] Mrs. Slaton, speaking the following day during a session for women, disparaged local control by suggesting that the nation's 140,000 local road officials constituted a waste of millions of dollars that would be better spent on expert officials and state-level administration.[28] Georgia senator Hoke Smith expressed support for both long-distance highways and recent federal legislation for farmers but stopped short of fusing the two in support of highway legislation.[29] Carl Fisher's friend Clarence Kenyon, president of the Indiana Good Roads Association, spoke in favor of federal highway aid but warned audience members that it would require tireless lobbying on their part. State and federal aid could only be obtained, he warned, "by getting next to the legislators, and going after them."[30]

Even Slaton's prominent role in both the meeting and the Dixie Highway announcement reflected a far-reaching optimism about what the highway could do for his state. With less than 6 percent of Georgia's 82,000 miles of dirt roads surfaced or even graded, Slaton had good reason to support an interstate highway proposal. But he also had good reason to try to divert attention away from Georgia's recent bad press. Over the past year, the national news media had descended on Georgia to cover the controversial trial of Leo Frank, a Jewish factory superintendent convicted of the murder of thirteen-year-old employee Mary Phagan in 1913. The scant evidence of Frank's guilt made headlines, exposing the inequalities of the state's justice system for all the country to see. No doubt Governor Slaton, whose own legacy was threatened by the Frank case, saw a modern highway initiative as an opportunity to redeem both the state's reputation and his own.[31] The Road Congress turned out to be the ideal backdrop for Fisher's highway proposal after all, though not only for the reasons he had predicted.

Carl Fisher was not the first to dream up a north-south highway, but he wielded enough influence to turn the idea into a crusade. His carefully orchestrated announcement took full advantage of the momentum of the Good Roads Movement in the South and the popularity of early marked trails. Fisher exploited the growing demand for state and federal highway aid even as he quietly promoted elite business interests like his own. Those interests would continue to guide the project even as governors and citizens took their places as constituents of this ambitious new interstate highway.

Dividing North and South

The proposal for the Dixie Highway came at a key moment not only in the history of the Good Roads Movement but in southern history, as Progressive legislation took up Populist themes aimed at helping farmers by expanding the reach of state and federal agencies. Although many southerners were still haunted by memories of federal occupation during Reconstruction, the dire state of agriculture in the South had long since begun to shift attitudes about government intervention. The Smith-Lever Act of 1914, sponsored by Georgia senator Hoke Smith and South Carolina congressman Asbury F. Lever, placed federally funded, state-supervised agriculture experts in almost every county in the United States to offer advice to farmers and promote more-scientific farming techniques such as crop diversification. Through home-demonstration agents, the Smith-Lever Act also reached into rural homes to promote better nutrition, teach new cooking and canning techniques, and even set up markets through which farm women could earn money by sewing, baking, and selling fresh produce and eggs. The Underwood Tariff of 1913 significantly lowered duties on many consumer goods that farm families needed or wanted, while increases in funding for post roads that same year eased the isolation of rural life.[32]

Yet rural southerners differentiated between useful and intrusive federal programs. As historian William Link has shown, farm families spurned moralistic and paternalistic reforms that challenged the social—and racial—status quo. Public-health campaigns such as hookworm eradication were popular, but implementation was difficult because southerners resisted the outside intervention and the higher taxes that such programs required. Rural parents were similarly wary of education activists who tried to impose compulsory education laws because farm families relied on the labor of every member of the household. Textile workers in nearby mill towns, many of them not far removed from the farm themselves, resented efforts to pass child-labor laws for the same reason. Unlike agricultural extension agents, who were often educated rural men and women and blended into the communities they served, labor organizers and education activists were outsiders who challenged the social and economic order in the rural South.[33] Thus while southerners became more comfortable with the role of state and federal government in their everyday lives, they refused to surrender to it altogether.

This understanding of the responsibilities and boundaries of state and federal intervention shaped local responses to the Dixie Highway proposal in the South. Following William Gilbreath's announcement in November 1914, a fierce contest over the highway's route developed in the southern states. Citizens along potential parts of the Dixie approved of state involvement but refused to allow the governors to dictate routes. Instead, they tried to assert their own influence on decisions about where the highway would go.

The initial proposal for the Dixie Highway did not specify precise plans for its course, but geography dictated the broad outlines of the route. Illinois, Indiana, Kentucky, Tennessee, Georgia, and Florida were on the most direct and obvious path from Chicago to Miami Beach. Major cities such as Indianapolis, Louisville, Nashville, Chattanooga, Atlanta, Macon, and Jacksonville made logical anchor points for a tourist highway and market route, as Gilbreath's original proposal had made clear.[34] But the exact course the highway would take between these cities was up for grabs. During the weeks leading up to the governors conference, dozens of routing contests developed among neighboring cities and towns in the Dixie Highway states. These contests were less intense in the North, where better roads and shorter distances made routing decisions more clear. But in the South, the stakes were much higher. The proposed routes in the South, far longer and more rugged, were composed of piecemeal roads that did not always join up. These remnants of spokes-on-a-wheel road systems presented problems that intensified the rivalries among small towns and rural counties. Moreover, in the South, the Dixie Highway would link many of these isolated communities to cities—and to one another—for the very first time. The ensuing debates over where to route it underscored not only vast differences between the economic and political resources in the North and South but also disparities within individual states, where rural and urban divisions often superseded regional ones.

While most roads in Illinois and Indiana were no better than those in the South, both states committed large sums of money to improving the main thoroughfares. The highway commission in Illinois was nearly a decade old in 1915, and state laws empowered it to finance one of the earliest systems of state-aid highways in the country. State funds assisted counties in the construction of many miles of standardized, paved highways under the supervision of trained engineers. Once completed, maintenance of these highways became the sole responsibility of the state.[35] Indiana did not have a highway commission, but it outpaced every other state in the nation in the value of local road bonds and had more surfaced roads than any other state. By 1915 Indianans enjoyed a travelable if limited network of main roads.[36]

When the Dixie Highway routing contest began in early 1915, Illinois and Indiana had the bare outlines of an interstate system already in place. Good roads in both states linked Chicago and Indianapolis, the two northernmost cities on the Dixie Highway, along a few different routes, but one of the best was an *L*-shaped path from Chicago 136 miles south to Danville, Illinois, and then due east toward Indianapolis. The first half of this route was the fine Vincennes Road, a former trading route through Indian Territory that became the first state highway—Illinois Route 1—in 1835, when the general assembly needed a road linking Chicago to the river town and military outpost of Vincennes, Indiana. By 1915 much of this route was paved. The other half of the route was an existing interstate road that linked the industrial city of Danville with Indianapolis. Citizens along a more-direct diagonal path between Chicago and Indianapolis vied for the highway as well, but their route passed through smaller towns and rural areas located almost exclusively in Indiana. Perhaps because they already had access to both cities, as well as nearby market towns, via existing roads, they did not put up a big fight when their neighbors in Illinois pushed for a longer route there.[37]

State aid influenced the routing competition south of Indianapolis as well, where several possible routes emerged but only two dominated debates. Most observers assumed the more-direct path through Louisville would win out over the alternative, which proposed an easterly detour through Cincinnati before turning south toward Lexington. Yet the latter route, while longer, had the advantage of adding a seventh state to the Dixie Highway, and one with a very active state highway commission. Ohio ranked a close second to Indiana in the value of local road bonds and the mileage of surfaced roads, and it led the nation in total mileage of paved roads. In fact, nearly half of the nation's paved roads were in Ohio. Kentucky's highway commission served only in an advisory capacity, but millions in county-road bonds had facilitated the completion of surfaced roads linking the state's two largest cities, Louisville and Lexington, north to Indianapolis and Cincinnati and to points farther south.[38] Nevertheless, the prospect of including Ohio on the Dixie Highway provoked the first real battle of the routing campaign, pitting Louisville to the east against Lexington to the west. The Cincinnati Chamber of Commerce launched a fierce campaign to route the highway due south through the bluegrass region because it bolstered Ohio's chances of being on the Dixie Highway. By early April, an observer described the contest between the Cincinnati-Lexington faction and its rival in Louisville alternately as a "hot fight" and "warm competition."[39] Despite the rivalry, however, residents along most potential routes from Illinois to Kentucky

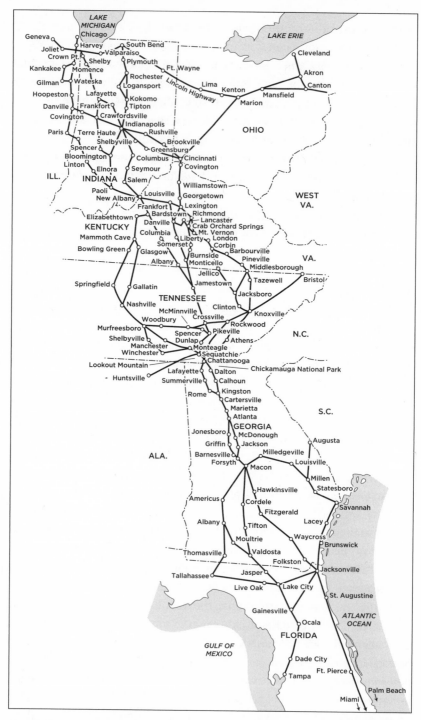

Map of all the competing routes for the Dixie Highway based on a hand-drawn map from the *Atlanta Constitution*, May 20, 1915, 1.

were assured either direct or indirect access to the Dixie Highway thanks to existing local and interstate networks. Accustomed to intercounty and even occasional interstate cooperation in road building, they found common cause, not just conflict, in the Dixie Highway.

Along with the availability of state aid, widespread assumptions about the primary purpose of the Dixie Highway also affected the intensity of the routing campaign as it moved south. The highway's midwestern organizers conceived Chicago as a starting point for a tourist route to Miami Beach and not simply as the northern terminus of a modern interstate highway between the two cities. Gilbreath's original highway proposal in November read like a complaint on behalf of northern tourists about the "lack of a thoroughly connected highway leading to the south," and subsequent press releases reinforced the understanding that the Dixie Highway was a route "leading down into the south."[40] While traffic would flow in both directions, routing decisions, like tourists, progressed from North to South.

This limited the potential for conflict in the North but multiplied it in the South, where routing decisions would be contingent upon the outcome of contests farther north. Just as the contest in Kentucky grew out of the campaign to add Ohio to the route, that rivalry in turn helped to shape pitched battles in Tennessee. The routing competition in Tennessee generated a web of possible routes, but three main options existed. One through the state capital of Nashville lined up with the proposed route through Louisville, while a route farther east through Knoxville aligned with the proposed route through Lexington. A third route in between those two passed through the rural communities due north of Chattanooga and would have linked up with either Louisville or Lexington. Thus Tennesseans along all three proposed routes took sides in the Kentucky contest and vice versa. This complicated the routing competition in both states but especially in Tennessee, where local governments controlled all of the state's roadwork. Although millions in county bonds had funded thousands of miles of surfaced roads, these roads were concentrated in the more heavily populated areas of middle Tennessee, near Nashville, while many roads in the mountainous eastern region near Knoxville were impassable.[41] The routing contest magnified these differences. Influential citizens along the Nashville route had the resources to establish a temporary organization—complete with printed stationary—to lobby the governor for the middle Tennessee route, which they said would serve "a great deal more people." According to the group's spokesman, Nashville and the surrounding counties also had enough money in the bank to complete the route.[42] The third route straight to Chattanooga appealed to supporters on both sides of the Kentucky rivalry, but it passed through sparsely populated counties that were largely inaccessible by road. While a fifty-mile detour through Nashville added time and expense to the Dixie Highway, both of the easterly routes would require more new construction.[43] With the outcome of the Kentucky contest still unclear and a great deal at stake along all three options, the routing competition in Tennessee grew more intense as the April governors conference drew near.

The cumulative effect of routing debates moving south along the Dixie Highway came to a head in Georgia, where competition gave way to more than a half dozen complex and spirited battles. Resolving these contests would be especially critical to the success of the highway. The gateway to Florida's vacation paradise, Georgia was the second largest of all the Dixie Highway states and would comprise a significant portion of the highway's total mileage. It was also the most heavily populated southern state, with over 2.5 million people clustered around some of the main cities and towns vying for the route. Atlanta, the state capital and symbolic capital of the New South, served as the region's undisputed commercial center. Along with Macon, eighty-five miles to the south, the Atlanta area was already a significant transportation hub for the region's main railroads.[44]

Part of the reason that the routing competition in Georgia became so intense was that the roads there were among the worst on the Dixie Highway, and it would be expensive to transform them into modern routes that could withstand heavy automobile traffic. A jumble of short, unimproved roads controlled entirely by 147 county governments comprised the state's primitive system of roadways. Road experts estimated the bare minimum cost of "improving" a road, which meant grading and surfacing a worn path into something resembling a proper roadway, to be $1,000 to $2,000 per mile, though it was often much more than that, while hard-surfacing cost several times that amount. Yet the average Georgia county in 1914 had 600 miles of meandering, disconnected roads and enough tax revenue to spend only $300 to $600 per mile.[45] Federal post-road funding supplemented some of the most important mileage in each county, but in 1914 Georgia received enough to give each county only $2.25 per mile. With such a dearth of funds, counties could afford to surface only a few roads with the cheapest and most abundant material available: a fragile mixture of sand and clay. Counties typically improved between twenty and sixty miles a year and struggled with the high costs of maintaining dirt roads.[46]

While most counties in Georgia faced the same financial challenges when it came to roads, regional differences within the state distinguished the various routing competitions. Atlanta and Macon anchored a three-part contest. To the north, the small cities of Rome and Dalton competed for the Dixie Highway between Chattanooga and Atlanta.[47] Along the short stretch between Atlanta and Macon near the center of the state, a contest developed between two parallel routes just a few miles apart. And in south Georgia, in the large, sparsely populated agricultural region known as the Black Belt (so named for its historic attachment to plantation agriculture), numerous routing options created one of the most complicated contests of all. Unlike in the North, no exemplary options existed in Georgia. Instead, a number of possibilities surfaced, each one having as many drawbacks as advantages and each becoming the focus of a fevered campaign through which local Georgians expressed both excitement and deep-seated anxiety about plans for the Dixie Highway.

The first major rivalry emerged in the northern part of the state between Rome and Dalton. Both cities were situated in a wide valley of fertile farmland between two ranges of the southern Appalachian Mountains, a critical passageway exploited for centuries by everyone from preindustrial travelers to local and long-distance railroad companies. Although they remained the two largest railroad communities in north Georgia in 1915, Rome and Dalton were still small towns by even regional standards. Rome had double the population of her sister city just thirty miles to the east, but Dalton straddled a newly improved, federally funded post road between Chattanooga and Atlanta. Named the Johnston-Sherman Highway in 1911 to honor the Civil War generals who fought along the same path in 1864, the road bolstered Dalton's claim to the Dixie Highway. Rome was a river city that remained dependent on steamboats and railroads, but it was still a vital transportation hub for farmers in the tristate area of Alabama, Georgia, and Tennessee. It was nevertheless isolated from nearer towns and cities by the condition of local roads. Once cars and highways surpassed railroads and waterways as vital arteries, cities like Rome were done for.[48]

Supporters of the Dalton route drew on years of experience in their campaign for the Dixie Highway. Since 1903 north Georgians had been

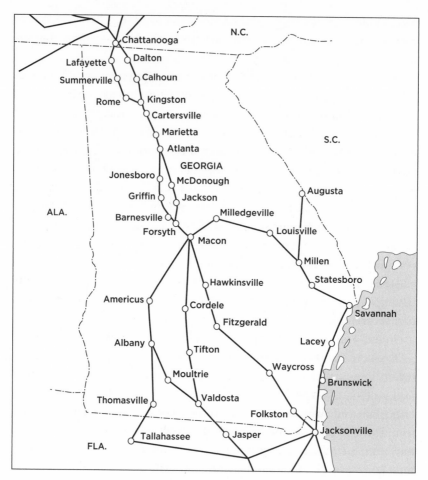

Competing routes for the Dixie Highway in Georgia based on a hand-drawn map from the *Atlanta Constitution*, May 20, 1915, 1.

campaigning for a highway between Chattanooga and Atlanta. In 1911 five neighboring counties between Dalton and Atlanta had organized the Battle Field Route Association to lobby their congressmen for a federal post road appropriation in order to build it.[49] By 1915 the Johnston-Sherman post road linked several famous battle sites between the Chickamauga National Military Park near Chattanooga, one of several Civil War battlefield parks established by Congress in the 1890s, and Atlanta. During the Dixie Highway routing campaign, citizens along the route often eschewed the formal "Johnston-Sherman" name in favor of the more-familiar "Battlefield Route" in order to commemorate popular Civil War battle sites along the route, such as Chickamauga and Ringgold Gap. Surviving Confederate veterans and the United Daughters of the Confederacy also joined the campaign for the Battlefield Route, which was to them a fitting memorial to the Lost Cause.[50]

Many Atlantans supported the Battlefield Route over the highway through Rome as well. Four years earlier, Fulton County (Atlanta) and heavily populated Cobb County just to the north had joined the Battle Field Route Association in order to petition for post-road funding. By 1915 their excellent city streets and intercounty post roads linked a large urban population to Dalton.[51] But Atlanta-area businessmen also wanted to exploit Civil War tourism, and the highway from Dalton linked historic battle sites in north Georgia to several nearer Atlanta, including Kennesaw Mountain. One of the most influential men in the Atlanta business community had a personal interest in the Battlefield Route as well. Ivan E. Allen was the founder of the Rotary Club and president of both the Atlanta Convention Bureau and the chamber of commerce, but he was also a Dalton native and son of one of the founding members of that city's chapter of the United Daughters of the Confederacy.[52] He used all of the considerable resources at his disposal to promote the Battlefield Route. On behalf of the Convention Bureau, he made a detailed log of the Battlefield Route and designed maps recasting the route as a 148-mile tour of historic Civil War sites all the way from Lookout Mountain in Chattanooga, the site of a critical Confederate defeat, to Stone Mountain just south of Atlanta, which would soon become the state's most famous memorial to the Lost Cause. The *Atlanta Constitution* all but endorsed the popular Battlefield Route early in the competition as well.[53]

The route through Rome was short on tourist attractions and support from powerful urban businessmen, so supporters highlighted the

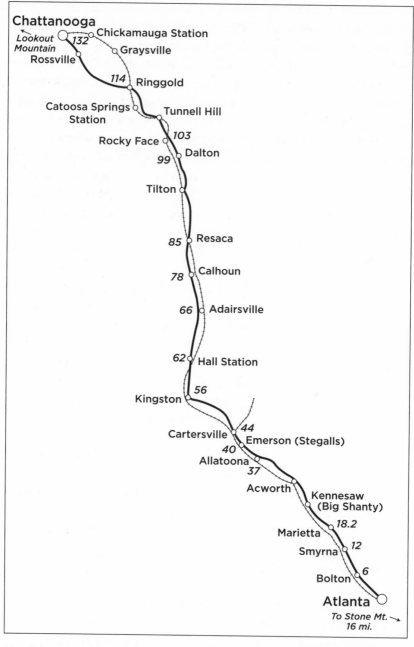

Map of the Battlefield Route based on a hand-drawn map from the *Atlanta Constitution*, March 28, 1915, 5F.

city's modern appeal as a growing market hub. Although their route was longer than the Battlefield Route, it served more full-time residents. No fewer than 69,000 people in the three northwest Georgia counties would straddle the highway through Rome, among them farmers taking crops to market, rural mail carriers delivering packages, and country merchants receiving shipments and customers. By the spring of 1915, the counties had invested hundreds of thousands of dollars to get the route into shape, and county representatives claimed that the entire stretch of the Dixie Highway between Chattanooga and Atlanta via Rome could be ready for dedication within one year. To compensate for the shortcomings of their indirect and unfinished route, they pointed out that Rome provided a "logical noonday stop" for tourists traveling from Chattanooga to Atlanta. Halfway between the two cities, if a little out of the way, Rome had plenty of fine restaurants, hotels, and garages to service travelers and their automobiles.[54] Romans were also investing hundreds of thousands of dollars in private capital and city bonds to build new tourist facilities and municipal buildings. The Battlefield Route may have given Dalton a more obvious advantage as a tourist highway, but residents near Rome believed that local resources and enthusiasm for the Dixie Highway more than made up for the shortcomings of their route.[55]

Although supporters of the Dalton and Rome routes emphasized their differences, citizens and local officials in both towns faced similar challenges when it came to completing their routes. The best sections of the Battlefield Route were contained within Whitfield County. And while Whitfield and Dalton officials boasted of having tens of thousands of dollars in the bank to continue improving local roads, smaller and poorer counties along the route struggled to do their share of the work. The sparsely populated Catoosa County to the north still had not completed its short eight-mile section of the route, and Gordon County to the south was in danger of never completing its section. Four years earlier, when neighboring counties were seeking the post-road appropriation to complete the Battlefield Route, Gordon County had been unable to raise matching funds in order to complete its portion of the road. In 1915 it remained the weakest link on the highway. Like many Georgia counties, Gordon County was still divided into colonial-era militia districts, but local tax laws forbade spending revenue raised in one militia district in another. This limited the road funds available to the county as a whole, making it difficult to concentrate funds on a single road through multiple districts, much less to coordinate roadwork across county lines. There was little Dalton city officials could do about what they called their neighbor's "embarrassing" predicament except wait until Gordon County officials changed local laws.[56] The roads in Rome and Floyd County were similarly unrepresentative of those in the two other counties along that route. While Rome-Floyd officials could afford to hire contract labor to work on their portions of the Dixie Highway, their neighbors struggled even to maintain their roads.[57] In January 1915 Chattooga County officials just north of Rome reported local roads to be in the worst condition in years, yet local voters, wary of taking on that much debt, defeated a $150,000 road bond referendum to improve them. Instead, the county sponsored a "Good Roads Day" in order to persuade local men to donate an extra day of work to the county's roads, even though the statute-labor system had long since proven to be unpopular and ineffective.[58]

Challenges along the Rome and Dalton routes helped to explain what made the competition so fevered. Neither had as obvious an advantage over the other as they claimed. While counties along both routes pledged to build modern bridges, hard-surface local roads, and complete their portions of the route in record time, their goals were wildly disproportionate to the condition of existing roads and local resources. Yet these promises were in direct proportion to what was at stake for the tens of thousands of ordinary citizens in the rural villages and small towns along both routes. Sandwiched between two mountain ranges and isolated from the rest of the country except by railroad, north Georgians staked their futures on the Dixie Highway.

Moving south along the highway from Atlanta to Macon, a second competition emerged between the small towns of Jonesboro and McDonough. The two towns anchored short parallel routes scarcely twenty miles apart. On the map, this competition resembled the Dalton-Rome contest, but in practice it reflected very different local concerns. While the competing routes in north Georgia spanned just three counties each, both middle Georgia routes paralleled heavily trafficked railroad lines between Atlanta and Macon. Each was only ninety miles long but cluttered with depot towns. On the Jonesboro route to the west, five small counties and ten depot towns along the Central Railroad line launched a determined campaign for the route. One observer described the route as so "thickly populated" with "so many small towns" that there was "not a stretch of the road two hundred yards long upon which there is not one or more houses." Along the McDonough route to the east, an almost equal number of depot towns served the Southern Railroad. Each route claimed to be the more direct and heavily populated one and to have the best sites and scenery for tourists, but all such comparisons ended in a draw.[59] The routes were nearly identical in every way.

The competition between such similar adjacent routes underscored the pervasiveness and the limitations of spokes-on-a-wheel road building in middle Georgia. Major railroads linked all of the towns along both routes to two of the largest cities in the state and points beyond, but good roads did not. Like the railroads it would parallel, the Dixie Highway would connect towns otherwise isolated by short feeder roads. Towns ten or twenty miles away from the route would remain dependent upon railroads, while their neighbors closer to the highway enjoyed automobile access to points farther north and south. With few roads that led beyond the nearest depot, routing the Dixie Highway through the heart of Georgia's railroad network accentuated the problems facing small railroad towns in an era of growing rural automobility.

As routing projections moved into the southernmost part of the state, the conflict exploded into the third and most complicated contest of them all. All of the difficulties of planning the Dixie Highway farther north accumulated in south Georgia's expansive Black Belt, one of the least developed regions on the entire route, the heart of the state's agricultural economy, and a major player in regional cotton production. Citizens proposed several different routes for the section of the Dixie Highway between Macon and Jacksonville, Florida, any one of which would reach only a minority of the state's farmers or local businesses. Early in the contest, three emerged as the main contenders, each with unique strengths but different objectives. The "Old Capital Route" was conceived as a tourist highway through historic former state capitals and the old port city of Savannah. The two others passed through the southwestern and southeastern regions of south Georgia, the former winding through farming communities and the latter a direct path through the vast Okefenokee Swamp. The southwestern route was easier to complete but out of the way, while the southeastern option required expensive bridges and upgrades to existing roads but was the most direct. The Old Capital Route was neither the longest nor the most expensive option, but it did bypass most of interior south Georgia. These three options for the Dixie Highway in south Georgia revived debates over building tourist highways at the

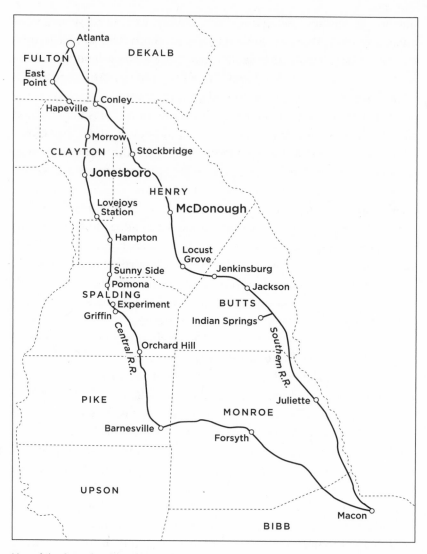

Map of the Central and Southern Railroad lines between Atlanta and Macon, Georgia.

expense of market roads and vice versa, but the stark differences among them also made it clear that no single route could meet south Georgia's needs, much less those of the state, region, or nation.[60]

Tension between the old and the new characterized the contest to place the port city of Savannah on the Dixie Highway. Once a bustling Old South commercial center and the state's first capital, Savannah suffered after interior railroad cities such as Atlanta surpassed it in the second half of the nineteenth century.[61] By 1910 Atlanta already had more than twice the population of Savannah. Even smaller railroad towns like Macon promised to outgrow the old port city in another generation or two.[62] Two other former state capitals on the route, Milledgeville and Louisville, struggled even more as the political and commercial centers of the state gradually moved inland. The river town of Louisville, 100 miles west of Savannah, had been selected to take over as the state capital in the late eighteenth century because it sat at the junction of three market roads to the major ports of Savannah, Augusta, and Georgetown. But within a decade, malarial outbreaks and westward migration prompted state leaders to move the capital even farther inland to Milledgeville.[63] Near the geographic center of the state, Milledgeville remained the capital long enough to grow from a frontier outpost to a sizable market center, but it could not keep pace with the railroad boomtown of Atlanta. Just three years after the end of the Civil War, the capital moved west once again to Atlanta, along with much of the business that had kept Milledgeville afloat.[64] As populations in and around the old capitals declined, so did local tax revenues. The advent of railroads only further compromised the area's roads. By 1915 roads in the region had suffered from decades of neglect.[65] Survival would depend on access to modern transportation routes, a lesson all three former capital cities had learned the hard way during the railroad boom.

Supporters of the Old Capital Route envisioned it as a tourist highway that would revive these fading jewels of colonial and antebellum settlement. While some proposed routing the Dixie Highway to Savannah via Macon, many preferred a link from Atlanta to Savannah that bypassed the upstart railroad hub altogether. This latter route followed the main path of Sherman's March to the Sea. Residents who wanted to cash in on Civil War tourism promoted the route as a natural extension of the Battlefield Route from Dalton through Atlanta.[66] It was a strategic alliance that could only help their chances of being selected, even if all they had to offer tourists was a crumbling antebellum mansion that supposedly had been spared because Sherman thought it too beautiful to destroy.[67] Eager to bridge their region's past glory with the promise of the New South, Savannah officials helped fund road improvements in other poorer counties along the route in hopes of making theirs the path of the Dixie Highway through south Georgia.[68]

The two other routes through the rural Black Belt could not have been more different from the Old Capital Route or from one another.[69] The southwestern route passed through some of the newest and most sparsely populated rural counties in the state. This region of the state was one of the last to be settled by Georgians migrating into former Creek Indian territory, but by the mid-nineteenth century, it was one of the wealthiest agricultural regions in the South. Cotton acreage in Georgia more than doubled between 1880 and 1914, while yields increased exponentially. In southwest and south-central Georgia, cotton production increased by as much as 1,000 percent in some counties during the boom years between the end of the Civil War and the turn of the twentieth century. Parallel increases in the worldwide market caused cotton prices to drop to all-time lows by the 1890s before shooting upward again during the First World War. Between 1900 and 1916, Georgia cotton-crop values tripled. After the war, the price of cotton plummeted again, from thirty-five cents a pound in 1919 to just seventeen cents a pound a year later.[70] The untimely arrival of the destructive boll weevil finally drove many Georgia cotton farmers into foreclosure by the early 1920s.[71]

The cotton monoculture not only left southwest Georgia farmers vulnerable to price fluctuations and natural disasters, but it also reinforced a railroad-based marketing system that isolated the region and forestalled both road construction and agricultural diversification.[72] Tiny depot towns along the Southwestern Railroad and the Muscogee Railroad served the primary purpose of linking cotton farms to major cotton markets. And local roads served the same system by feeding millions of bales of cotton into a saturated market.[73] One of the main reasons that the U.S. Department of Agriculture's crop-diversification crusade did not succeed for decades was the failure to understand that farmers in places like southwest Georgia needed better markets and market access—distribution networks—before they could diversify on a large scale. As Dan Hughes, a Georgia farmer and assistant commissioner of agriculture, argued in 1914, farmers could not diversify until businessmen did. "Give the farmer an outlet for diversified products," Hughes implored them, "and he will grow more . . . of them every year."[74] Good roads were a critical missing piece in this puzzle, as successful truck farmers in the Northeast, with its large urban centers and good roads, had already discovered.[75]

The proposed Dixie Highway route through southwest Georgia filled an important void in the region's agricultural marketing system, and at just the right time. Although cotton farming showed no signs of giving way altogether, by 1915, thanks to northern investors, new crops were poised to challenge cotton for a small share of the regional economy, none more so than peaches and pecans. Peach and pecan orchards had expanded rapidly since 1900. That year, the value of all fruit and nut crops in Georgia totaled just $672,000, while cotton generated over $57 million in profits. But by 1915 the state department of agriculture estimated that peaches alone brought in $3.5 to $5 million. Most pecan orchards had not yet matured, but farmers had already invested between $10 million and $15 million in orchards that were yielding as much as $1,000 to $1,500 per acre.[76]

Emerging market towns for these new crops anchored the proposed Dixie Highway route moving south from Macon. The first of these towns was Fort Valley, the center of the state's peach industry. Peach production had taken off there in the 1890s and was booming by 1915, but even the advent of refrigerated railroad cars—and a state law requiring railroads to furnish them for peach and melon shipments—and more flexible freight schedules did not make railroads an ideal means of shipping fragile and perishable fruit crops.[77] South of the peach farms, good soil, railroads, and a prime location along the Flint River had made Sumter County one of the wealthiest counties in the cotton-producing Black Belt. By 1900 it was also home to several sawmills and turpentine distilleries, and the county seat of Americus was one of the largest and most important market towns in southwest Georgia.[78] Forty miles south of Americus, the town of Albany was the center of the new pecan industry in Dougherty County. The first pecan orchards had been planted there years earlier, and by 1915 the trees were plentiful and mature enough to support a thriving industry. That same year, the Georgia-Florida Pecan Growers Association and the National Pecan Growers Association pooled their resources in order to establish a marketing cooperative in Albany. Along with the Fruit Growers

Two men driving a Buick along the Dixie Highway near Albany, Georgia, 1915.
(Courtesy of the Georgia Department of Archives and History, Vanishing Georgia
Collection, Image DGH-62)

Exchange in Atlanta, they helped diversified growers market their crops
as efficiently as possible and, in the case of fruit growers, often intervened
when railroads damaged the fragile crops.[79]

While market access was of utmost concern to peach and pecan growers, they also recognized the potential for automobile tourism. Boosters
of the route argued that acres and acres of peach orchards and pecan
groves offered tourists "an unusual diversification of scenery and crops"
that they would not find elsewhere along the Dixie Highway. "Just at this
time of year," said one observer in the spring of 1915, "when the return
traffic from Florida would be moving, there is no sight in the country
more charming than the blossoming peach orchards all the way from
Americus to Macon. Traversed by such a touring route, peach blossom
time in Georgia would soon be as celebrated as cherry blossom season
in Japan."[80] Near Americus, the Andersonville National Cemetery drew
tourists who were interested in the Civil War. The president of the Albany
District Pecan Exchange argued that the "tremendous" and "magnificent"
pecan industry would be of particular interest to midwestern tourists be-

cause many of them were investors in local pecan groves.[81] And near the southernmost end of the route, along the Georgia-Florida line, the city of Thomasville was already a winter resort for wealthy northern tourists seeking warmer weather and the solitude of the longleaf pine forests.[82]

Northern investors and growers joined together with ordinary citizens and farmers to campaign for new agricultural and commercial opportunities for southwest Georgia along the Dixie Highway. Not only would it assist the region's largest market towns by facilitating crop diversification; it also would provide vital highway access to an entire region of the state isolated by bad roads. This combined support for the Dixie Highway was especially important because the route through southwest Georgia was neither heavily populated nor the most direct path to Miami Beach. Enthusiastic citizens prevailed upon local officials to make improvements to local roads in time for the route selection. Sumter County officials redoubled their efforts in order to complete their thirty-five-mile section of the route by early April.[83] But neighboring communities eager to attach themselves to the Dixie Highway also lent a hand. In Columbus, a fast-growing textile-mill town on the Chattahoochee River sixty miles west of the proposed route, local officials announced plans to complete a road to Americus. They even agreed to help a smaller adjacent county complete its portion of the road.[84] Ordinary citizens also came together to demonstrate their support for the highway. In the tiny town of Camilla, at least 500 people from surrounding counties turned out for a Dixie Highway meeting, and two days later, a meeting in Americus drew 650.[85] A week later, people from all the counties on the route held yet another "splendidly enthusiastic" meeting to endorse their collective effort to give southwest Georgia a market route and modern interstate highway.[86]

Agricultural diversification also figured into the rival routing campaign through southeastern Georgia, but only because the economic advantages of diversification were waning by 1915. The extension of railroad lines into the region after the Civil War had helped cotton farmers seeking new markets for their cotton, but above all else, it had allowed the timber industry to prosper. Between 1875 and 1900, timber and turpentine generated fortunes and doubled the population of some wiregrass counties in the span of a generation, but by 1915 the boom was nearly over. Most of the longleaf pines in the area had been logged, glutting the lumber market, driving down prices, and eliminating jobs and fortunes. Landowners began planting a new fast-growing variety of pine trees to replace the longleafs, but it would take years for them to mature. In the meantime, farmers weathered the ups and downs of the cotton market, while thousands of others left the region for jobs in nearby cities like Jacksonville, Florida.[87]

Fortunately for southeast Georgians, geography gave their flat, piney-woods landscape a significant advantage over the proposed route through southwest Georgia's peach and pecan orchards. The Short Line Route was a direct path that was seventy-five miles shorter than any other proposed route between Macon and Jacksonville, but it was longer within Georgia by over 200 miles, spanning nearly two-thirds of the length of the state. The route also passed through half a dozen counties and numerous small towns. By the spring of 1915, these communities had pooled their resources in a determined effort to link local roads into a highway of uniform size and quality.[88]

Citizens along the Short Line Route understood that the rivers and the Okefenokee Swamp—a vast 438,000-acre wetland straddling the Georgia-Florida border—were attractive to tourists and hunters, and they were eager to invest in a modern highway that would repay them in tourism dollars. That meant building hard-surfaced roads, a rarity in southeast Georgia but practical and necessary for a heavily trafficked route through flat, sandy land. Critics of dirt roads in Ware County complained that the county wasted tens of thousands of dollars a year on roads "that do not last longer than the first hard rain that falls." Nearly all local roadwork was repair work. "We throw up a little sand and dirt—do a little grading," one critic said, "and then go to some other part of the county and do the same thing." After every rain, the cycle started all over again.[89] Residents of even small towns along the Short Route agreed that modern paved roads should replace dirt roads, and they tried to persuade others that they could raise the money to build them. The mayor of Ocilla, population 2,017, promised to build "a magnificent all-year-round road . . . whether or not we are on the route of the Dixie highway."[90] They did not control the state's agricultural wealth like their neighbors to the west, but local people were no less determined to prove themselves worthy of—and able to build—a long-distance highway. They showed up by the hundreds at meetings to boost the route. Early in the routing contest, they even established an official Short Route Association, much like the one along the Battlefield Route in north Georgia, to coordinate the work of the several county governments responsible for a share of the route.[91]

Small signs of progress appeared during the routing contest, as local officials cooperated across county boundaries to stitch rugged roads into something resembling a travelable highway. County road commissioners concentrated all of their resources along the main roads that would make up their sections of the Dixie Highway.[92] Officials in Ware County, one of the most populous counties on the route, supplemented the labor of chain gangs with modern road-building machinery to smooth and grade all of their "important" roads.[93] In less-populous Wilcox County, commissioners said they were "sparing neither time nor expense" to make their piece of the Dixie Highway "as good as the best of it" by widening and packing the dirt roads and installing steel bridges and culverts.[94] Along with these carefully coordinated efforts to improve and link up main roads, the Short Route's geographical advantage gave citizens in southeast Georgia good reason to believe they would win the routing contest.

Geography, however, also proved to be the Short Route's greatest disadvantage. The proposed road passed right through the eastern edge of the Okefenokee Swamp. Railroads and logging companies had built lines to the edges of the swamp in order to harvest the cypress trees there, but even the mighty railroads had not attempted to cross the swamp.[95] The Short Route also crossed several rivers that flowed into the swamp, and the low-lying, sandy roads in the region flooded badly when it rained.[96] The cost of building new bridges combined with the high maintenance costs neutralized the Short Route's main advantage over longer rival routes. With such great challenges and limited resources, wiregrass communities struggled to maintain their existing roads, much less build new ones.

Shorter distances and big promises were no match for the challenges of building a modern highway through a sparsely populated region dominated by sandy soil and wetlands. With local populations in decline, tax revenues struggled to keep pace with the demands of modern highway construction. Wilcox County, for instance, had promised to build a fine road but near the end of the routing competition had only completed half of the necessary work.[97] Months after singing the praises of permanent roads, Ware County officials were still pouring money into repairing dirt roads after the rainy winter season, and construction on a critical bridge over the St. Marys River had not even been started.[98] Without the bridge, the Short Line Route was impassable. In order to convince scouts for the Dixie Highway that they had a viable route, locals considered temporarily rerouting the southern half of the Short Line farther west in order to circumvent the Okefenokee Swamp and the St. Marys River. The additional mileage eroded the Short Line's main advantage over rival routes, but citizens desperate to be on the Dixie Highway preferred a short-term detour to a permanent bypass in favor of another route.[99]

The same problems plagued many communities along proposed routes for the Dixie Highway in the South. Still, some beleaguered participants in the routing contest saw a silver lining in the intense competition. Rivalries, said one observer, guaranteed "hundreds of miles of good roads, the construction of which would have been delayed many years" without the momentum of the Dixie Highway campaign.[100] Efforts to build feeder roads and alternate routes even before the designation of the official route vindicated the Dixie Highway's challenge to the old spokes-on-a-wheel model of road building and created communal identities across county lines. By cooperating with neighboring counties, local governments surrendered total control over their own roads. Preoccupation with local needs gave way to a common effort that served both local needs and the collective demand for better roads. Noticing the extent of road improvement, one observer remarked: "[N]ot a section of south Georgia from the coast to the Alabama line . . . is not either petitioning the commissioners to send the highway through their respective sections, or over such a course as will allow them to build roads connecting with the highway." Another observer predicted that the competition would generate an elaborate system of modern highways, regardless of which routes were selected for the Dixie Highway.[101]

■ ■ ■

But boosterish rhetoric could not conceal the long-term challenges that the routing contest unearthed. While organizers of the Dixie Highway had predicted that it would unify North and South, during the winter and spring of 1914–15, the routing competition only magnified existing regional divisions. The contest revealed vast differences in the local resources of the hundreds of communities along rival routes between Chicago and Miami. Regional divisions between the North and the South were most pronounced, but the heated contest in Georgia exposed the challenges of building a modern highway that suited the needs of diverse groups of citizens, especially those in regions with few good roads of any kind. Southerners may have been united in their transportation crisis, but they were divided over where and how to build roads. Local control over roads only multiplied the challenges of creating a modern transportation system that linked North to South, farms to cities, and local to national.

The routing contest was divisive, but it also revealed that farmers, urban businessmen, and everyone in between shared a commitment to, and a faith in, new transportation networks. The stakes were high for everyone in the routing contest, but southerners had to go to greater lengths to prove that their own roads were good enough to form the foundation of a modern interstate highway. The routing contest proved one main thing: getting a diverse public behind the Dixie Highway was easy, but pleasing them all would be impossible.

The Second Battle of Chattanooga

In April 1915 some 5,000 citizens traveled to the governors conference in Chattanooga from all over the Midwest and South in hopes of influencing routing decisions for the Dixie Highway.[102] Months of propaganda from Carl Fisher and William Gilbreath had given everyone the impression that the governors would select the official route, but only after hearing from delegates from each county along the proposed routes. Excited and anxious to learn whether or not their campaigns would pay off, farmers, businessmen, and county officials poured into Chattanooga to plead their cases one last time.[103]

Yet despite the intensity of the routing contest and the anticipation surrounding the governors conference, no one knew how or when routing decisions would be resolved. The governors had no authority to designate state, much less interstate, routes, and privately they expressed uncertainty about their roles in the selection process. Governor Slaton of Georgia responded hesitantly to inquiries about his intentions for Georgia, and on at least one occasion, he suggested that the decision should be made by engineers rather than politicians.[104] The role the states would ultimately play in helping to complete the highway also remained unclear. They could pledge only their support and cooperation, nothing more tangible. Even the state highway commissions in Indiana, Ohio, and Kentucky had limited powers within state lines and none beyond them. Local governments had the authority to build roads, but the routing contest had proven that they could not agree on an interstate route, much less build one, without help. So even as citizens had looked to their local officials to get local roads into shape for the routing decision, local officials looked to a higher power to coordinate work on the Dixie Highway. They found it only nominally in the Dixie Highway Association, established on the eve of the conference by Carl Fisher and a group of wealthy businessmen from the Dixie Highway states who signed the organization's charter and donated $1,000 apiece to become its "founders." Precisely how the fledgling organization would function remained to be seen, but the founders did not plan to replicate the role of the Lincoln Highway Association's leaders. They would provide funding for surveys, maps, and promotional materials, but public authorities alone would have to build the road.[105] All of the uncertainty only added to the nervous enthusiasm in Chattanooga. By the time the conference began on April 3, everyone wanted to be involved in planning the Dixie Highway, but no one knew how to do it.

The routing competition had been fierce, but the governors conference in Chattanooga proved to be the ultimate battleground, not only for rival routes but also for competing ideas about how to build a public interstate highway. It was the first time Dixie Highway organizers, founders, state leaders, and citizens met face-to-face to discuss the business of building the highway, and it did not go well. The afternoon session on April 3, described as "stormy" by one reporter, revealed mutual distrust among the governors, citizens, and DHA founders. Not surprisingly, the greatest friction centered on the routing decision. First, the governors tried to choose the route themselves, but the founders threatened to withdraw their financial support of the DHA unless they had a vote. The governors relented and agreed to cooperate with the founders, only to be forced to change their plans when the audience objected—loudly—to the fact that the names of all the founders had not yet been made public. When conference organizers revealed the identities of the founders, they read the names so quickly that the convention's stenographer complained he had been unable to "catch them all." Subsequent efforts to give them a voice in the route selection drew even more protests, this time led by Governor Ralston of Indiana and Governor McCreary of Kentucky.[106] William Gilbreath was unable to deliver the address he had prepared and stood by helplessly as the conference deteriorated into a fight. "The good roads movement is not local nor is it sectional," he had planned to say.[107] The scene before him seemed to contradict that and everything else the Dixie Highway was supposed to represent.

The blow-by-blow accounts of the raucous debates that filled the pages of the next day's local, regional, and even national newspapers conveyed deep suspicions about the motives of the wealthy founders.[108] These men were not mere do-gooders or philanthropists but savvy businessmen with significant political influence, as useful to the DHA as the DHA was to their businesses. John A. Patten owned the Chattanooga Medicine Company. Skirting the Pure Food and Drug Act of 1906, Patten's company still sold patent medicines such as Black Drought laxative and the fabled Wine of Cardui, whose 19 percent alcohol content probably did relieve "female troubles," and most other ailments, at least temporarily. A highway through his hometown could only boost business. The same was true for fellow Chattanoogans Charles E. "Charlie" James, a real estate developer; Claude H. Huston, a banker; and Richard Hardy, whose Dixie Portland Cement paved many miles of roads in the days before asphalt.[109] Kentucky-born chemical magnate Coleman DuPont, a longtime supporter of the Good Roads Movement, shared Carl Fisher's interest in Florida as well. He and his family invested heavily in Florida real estate in the 1910s, and DuPont often cavorted with Fisher at the exclusive Cocolobo Club in Miami Beach, where capitalists and politicians rubbed shoulders with celebrities such as Will Rogers and prizefighter Jack Dempsey. Like his friend Fisher, DuPont had every reason to believe that his small investment in the Dixie Highway in 1915 would pay off in a big way.[110]

Suspicion about the founders inflamed tensions, but the governors conference erupted because the stakes were high for everyone involved. The DHA's founders viewed the Dixie Highway as a political project that would benefit them personally and financially. Governors and county officials were under pressure from their constituents to support the highway. And citizens who could afford to travel to Chattanooga for the governors conference were not representative of the southern rural majority, but their apprehension about who would control critical decisions about the Dixie Highway typified a region desperate for better roads but wary of outside interference. Enthusiastic southerners joined automobile men and their powerful business allies to promote the Dixie Highway, but they distrusted them to make decisions that could make or break local communities.

Just one month earlier, a Georgian anticipating the governors conference had predicted it would be "the biggest real-in-earnest-do-something good roads meeting ever held in the South," but before it was over it had become known as the "Second Battle of Chattanooga."[111] The uproar over the founders threw the meeting into chaos and forestalled any further attempts to resolve the routing dispute. Unable to reconcile their differences, the conference-goers reached a compromise whereby neither the governors nor the founders would be in charge of the route selection. Instead, a commission consisting of two "impartial representatives" appointed by the governors of each state would be responsible for reviewing all of the competing routing proposals and selecting the official route in May. The board of directors was not an entirely new idea: for weeks leading up to the conference, rumors swirled that the governors might delegate all or part of the routing selection process to some kind of impartial committee, but the fight in Chattanooga made it a necessity.[112] Such an anticlimactic end to an animated meeting reflected the great difficulties of planning an interstate, interregional route through the road-poor South.

■ ■ ■

The DHA's board of directors was supposed to serve as a hedge against both the unilateral decision making that had hurt the Lincoln Highway Association and the pecuniary and political prejudices that had threatened to undermine the Dixie Highway during the governors conference. But despite the DHA's emphasis on impartiality, the directors appointed by governors of the Dixie Highway states included businessmen who would benefit directly from the highway's completion. Kentucky director Harry B. Hanger was a contractor from Richmond who built major infrastructure projects up and down the East Coast. Indiana director Thomas Taggart was a former Indianapolis mayor and close associate of both Governor Ralston and Carl Fisher. By 1915 he was also the owner of a resort hotel at French Lick Springs, just a few miles from one of the proposed routes between Indianapolis and Louisville. Even more conspicuous was Governor Ralston's selection of Fisher himself as Indiana's second director. Tennessee governor Tom Rye denied requests to appoint DHA founder Charlie James, diplomatically citing his obligation to select men with "no interest in any particular road." But both of his choices, Judge Michael M. Allison of Chattanooga and Colonel A. M. Shook of Nashville, were prominent citizens on two rival routes. The newspaper editors who served the DHA represented major cities and towns that would almost certainly be included on the Dixie Highway route, including Richard J.

Finnegan of the *Chicago Times* and Georgians William T. Anderson of the *Macon Telegraph* and Clark Howell of the *Atlanta Constitution*.[113] At the very least, appointing men from rival routes assured some balance in the decision-making process. In addition, a resolution adopted at the governors conference explicitly stated that the work of the DHA's board of directors could not conflict with the work of "duly constituted" county or state road officials. It was a reminder of the limits of the DHA's powers.[114]

Those limits were tested just days after the governors conference, when a power struggle erupted over control of the organization. On April 8 founder Charlie James usurped the authority of the board of directors by announcing on the front page of the *Chattanooga Times* that the DHA would route a portion of the highway through Signal Mountain near Chattanooga. By making the announcement so soon after the governors conference, James took full advantage of the DHA's vulnerability as a new organization. The DHA had not yet elected any officers, but James claimed to be its "president" and spoke on its behalf. Arguing that the abrupt routing decision was just the kind of "heroic treatment" the Dixie Highway's supporters wanted, James accused the governors of making appointments to the board of directors in exchange for political favors. The newspaper took James's side, portraying the DHA as a bulwark against the "petty politics and selfish politicians" that were manipulating the highway for political gain. Neither James nor the newspaper acknowledged that James owned an inn, a streetcar company, and 4,000 acres of undeveloped land on Signal Mountain.[115]

James tried to exploit the fledgling organization, but his brazen and transparent coup only strengthened the resolve of the DHA's supporters to adhere to the resolutions adopted during the governors conference. They were the only things holding together this fractious, fragile coalition of Dixie Highway supporters. Embarrassed by the controversy, the directors moved swiftly to rebut James and to remind everyone that they would be responsible for selecting a route that served the public good and not individual interests. In a letter circulated to all of the governors of the Dixie Highway states, director Michael Allison apologized for James's behavior and implored the governors not to allow the embarrassing episode to "lessen the enthusiasm for . . . this great national project." Citizens along rival routes also complained to the governors that James's "highhanded methods" constituted "a direct insult—a slap in the face—to every Governor who attended the Conference."[116] To his correspondents,

Governor Slaton of Georgia expressed his regret "that any friction should arise which would imperil the success of an undertaking so full of promised benefit to our respective States," but he deferred the authority of the board of directors in dealing with James. "The matter lies in your hands," he told director Clark Howell.[117]

Although powerful private interests continued to influence the process of building the Dixie Highway, the James controversy demonstrated that the DHA was up to the challenge. Under the direction of a state-sanctioned board of directors, the DHA developed the capacity to function as a viable quasi-public, pioneering highway-building agency. It took both civil and political influence to do so at a time when there was no federal agency empowered to build highways and no private organization that had succeeded. In a way, the DHA served as a stand-in for the federal highway agency that its supporters, and good roads advocates more broadly, had been demanding.

■ ■ ■

The governors conference and the James debacle profoundly changed the future of the Dixie Highway, but it had little impact on the intensity of the routing competition, which waged on for several more weeks. After the conference, however, the DHA gave greater focus and direction to the ongoing rivalries along the proposed routes. Three weeks after the conference, after the chaos surrounding it and the James controversy had died down, the DHA directors met to establish some basic guidelines for choosing the official route. They agreed to make their selection based on practical criteria such as mileage, condition of existing roads and bridges, local resources available for building or improving sections of the route, plans for maintaining those sections, the population each segment of the route would serve, and the number and quality of feeder roads. Aware of how difficult it would be for local governments, particularly those in the southern states, to do their share of the work, the directors wanted to identify the shortest route over the best existing roads in order to eliminate as many major upgrades and as much new construction as possible. They asked counties along the prospective routes to submit the required information to their state directors by May 10, which gave them little more than two weeks. The directors then would have only ten days to review the proposals before the scheduled May 20 routing meeting in Chattanooga.[118] Such demanding requirements narrowed the range of possible routes but did little to temper the rivalries. Supporters anticipated "a friendly contest, yet [one] fought determinedly" over the coming weeks and predicted a "lively scrap" at the meeting itself. One supporter expected yet "another battle of Chattanooga."[119]

It started out looking like these predictions would come true. On May 20, 1915, thousands of citizens poured into Chattanooga just as they had for the governors conference six weeks earlier, still intent on influencing the routing decision. The city of Rome sent 1,000 supporters up in a special eight-car train on the Central of Georgia Railroad, along with dozens of automobiles to prove that the route between Rome and Chattanooga was in drivable condition. Not to be outdone, Dalton sent a train and a caravan of fifty cars, and city officials pronounced May 20 a city holiday. Little did they know that the directors arrived in Chattanooga still uncertain about which routes to select in Georgia or anywhere else. Over the previous few weeks, they had toured many of the proposed routes and seen how fruitful the routing competition had been. County governments and local citizens had shown tremendous flexibility and adaptability in their efforts to secure spots on the highway. Their roads, though far from complete, had undergone major improvements, and county officials had demonstrated their understanding of what it would take to build inter-county roads and interstate highways.[120] Yet among major competitors like Dalton and Rome, no single option appeared significantly better than the other. Selecting a route was no easier for the DHA directors than it had been for the governors six weeks earlier.

The May meeting, however, proved to be nothing like the governors conference. No fights broke out, and no one suggested postponing the route selection. Instead, after two days of careful consideration of the merits of all of the possible routes, the DHA directors reached an extraordinary solution. The competition among rival routes in the South inspired Carl Fisher to suggest a "ring route" that circled the Midwest and the South all the way from the Great Lakes to the Gulf of Mexico, guaranteeing spots on the highway for all the major contenders. Georgia director William T. Anderson presented the official motion to designate parallel eastern and western divisions of the Dixie Highway "where it appears advantageous to do so" rather than to resolve any particular rivalry. The directors also voted to extend the Dixie north to the Canadian border at Sault Ste. Marie, Michigan; to allow official feeder roads; and to prohibit toll charges on the highway. The directors failed to designate just

Dalton citizens who traveled to Chattanooga for the routing decision in 1915. On the bumper, left to right, are Frank S. Pruden, B. A. Tyler, and a Mr. Bishop of Adairsville. At far left, Tom Boaz of Calhoun has his right foot on the ground. In the back are J. G. McLellan; Dr. J. L. Jarvis (bow tie); Louis Crawford; Col. J. J. Copeland; and Paul B. Fite, who is holding a trophy they were given for having more cars in their caravan than Rome-route supporters. (Courtesy of the Whitfield-Murray Historical Society)

two portions of the 6,000-mile highway system—the western division in Florida and the eastern division through southeast Georgia—but agreed to select those routes within ninety days.[121]

The bitterest contests in the South evaporated with the approval of Anderson's motion. Tourists could lunch in Rome on their way south and take in the charms of Dalton on their way back, or they could travel parts of both the eastern and western routes by using any one of the several east-west roads that linked the two divisions of the highway near major cities. Farmers also benefited through more options for marketing their crops. Some of the tensions between businessmen who wanted a tourist highway connecting major cities and farmers who needed a more-practical system of local and long-distance roads now seemed solvable.

In the spirit of cooperation that marked the routing meeting—and perhaps in an effort to smooth over tensions from several weeks earlier—the DHA elected Charlie James as its first president, and his chief critic and fellow Chattanoogan, director Michael M. Allison, became vice president. The new bylaws of the DHA stated that its purpose was to promote not only the construction of "a continuous improved highway" between the North and the South but also the building of other highways adjacent to the Dixie. They also established three levels of membership, both to encourage interest in the Dixie Highway and to ensure that operating expenses would be covered. New "founders" could join for a one-time fee of $1,000; councilors could secure their own lifetime membership for $100; and ordinary members could join for a $5 annual fee.[122] For the time being, it would remain for local governments to fund construction and maintenance of those highways, but they would have steadfast partners in one another and in the wealthy businessmen who financed and led the Dixie Highway Association.

■ ■ ■

The idea that the route could accommodate multiple pathways at once transformed the Dixie Highway from a single long-distance road into an interstate highway network, the first of its kind in the United States and the most significant transportation development in the South since the region's first railroads. Gone was any notion that this was merely a north-south tourist artery or market road. In its place emerged a plan for a standardized network of public roads and a toll-free interstate highway system.

Yet including multiple pathways also created new challenges. On June 1, just days after the Chattanooga meeting, Charlie James resigned as president of the Dixie Highway Association. James had never gotten over the decision to delegate to the directors what he thought could have been settled more efficiently during the governors conference back in April. Claiming to have spent days trying to make sense of all the routes designated at the May 20 meeting, James declared that the directors had made the highway "as complicated as it was possible to make it" and accused them of "cater[ing] to the interests" of too many supporters. It especially frustrated James that the directors had made the "arbitrary" decision to select eastern and western routes around the more-direct "short line" route through Kentucky and Tennessee. Though James never admitted it publicly, everyone knew the decision had cost him a fortune by bypassing his property on Signal Mountain.[123]

For the second time in as many months, James forced the DHA to reaffirm its commitment to the public interest. Georgia director Clark Howell used his own newspaper, the *Atlanta Constitution*, to publicly rebuke James and defend the route. The route through Signal Mountain would have required miles of new construction through undeveloped lands that were devoid even of railroad facilities. "[T]he committee acted upon the principle that the best existing roadways should be adopted for development," Howell explained, and he chided James for his poorly disguised effort at self-promotion. James was selected to be president, Howell said, "in the confidence that his general interest in the [Dixie Highway] movement would not be dampened by his loss of the central route, along which he has large property interests. It seems that we misjudged him."[124] Without hesitation, the DHA accepted James's resignation and appointed Michael M. Allison as his replacement.[125]

James was selfish, but he was right: the "ring route" had eliminated the routing contests but multiplied the difficulty of completing the highway, especially in the South, where the Dixie Highway now traversed more mileage than anywhere else. Construction of both eastern and western branches of the Dixie Highway, not to mention numerous official feeder roads, would require the cooperation of hundreds more county governments between Michigan and Miami Beach. The new longer route promised to serve millions of people traveling along the highway, but it also would affect the lives of hundreds of thousands more citizens living along its path. Adding to the challenge was the fact that the DHA gave each county just twelve months to complete its portion of the route.[126] If the prospect of a single North-to-South highway had inspired a fierce routing competition, building a far-more-extensive interstate highway system within a year would come with its own share of new trials.

Lobbying for Federal Aid

From the moment they announced the plan to build the Dixie Highway, Carl Fisher and William Gilbreath embraced the Good Roads Movement's campaign for federal highway aid. They envisioned the Dixie as a government highway—a "National project," as the founders described it—that would serve at once as a practical route for tourists and farmers and a testimonial for federal funding.[127] Citizens and local officials who participated in the routing competition also believed federal highway legislation would be forthcoming, an assumption that helps to explain their willingness to take on a project they could scarcely afford. If local officials started the Dixie Highway, they thought, the federal government surely would finish it.[128] The DHA nurtured these assumptions right up to the governors conference, where they passed a resolution promising to ask Congress to fund the Dixie Highway.[129]

The chaos surrounding the Dixie Highway routing competition and the route selection had temporarily eclipsed the broader campaign for federal highway aid, but challenges to the highway throughout the rest of 1915 and early 1916 brought it back into view. These challenges were more pronounced in rural areas of the South than elsewhere along the highway. DHA officials reported that by the summer of 1915, the 133 counties along the route had raised some $9,726,400 through taxes and county bonds. But nearly 90 percent of the total had come from urban counties in just three states: Illinois, Tennessee, and Florida. With just 123 miles of the Dixie Highway within its borders, Illinois had less mileage than any other state but raised more than a third of the total through large bond issues in just two counties. Georgia ranked third in total mileage but first in the number of counties the Dixie Highway traversed, yet the state reported raising just $159,000. Only Carl Fisher's home state of Indiana raised less.[130]

Despite the flurry of plans and promises that had immediately followed the routing announcement, most of the highway was far from complete at the end of the DHA's twelve-month deadline. It took nearly five months for Florida's directors to even designate the western division there.[131] Clark Howell and William Anderson took even longer to select a route through southeast Georgia because they could not agree on one. Anderson supported the Short Route through Waycross, while Howell backed the Old Capital Route through Milledgeville and Savannah. On March 25, 1916, the entire board of directors finally decided to adopt both routes, increasing Georgia's mileage but also its financial burden for the highway.[132] Slow progress in Tennessee and Georgia prompted the DHA to consider rerouting several portions of the highway. These were hardly idle threats: in July the directors stripped the eastern division of the highway between Atlanta and Macon via McDonough of its official designation after determining that "nothing has been done by a majority of the . . . counties to fulfill the pledges previously made to the association or meet the requirements of the Dixie Highway Association in providing a well surfaced and drained all year round road."[133]

The DHA's uncompromising standards only reaffirmed the need for federal highway legislation. Fashioning a thoroughly modern highway out of hundreds of meandering dirt roads required money and expertise; no amount of enthusiasm would make it easy for county governments in the South to build the Dixie Highway on their own. So even while they pressured county officials to speed up the pace of work all along the highway, DHA officials barnstormed the country in support of federal aid. Just days after the May 1915 routing meeting, founder Richard Hardy told a crowd in Atlanta that it was the federal government's responsibility to build and improve the nation's roads.[134] William Gilbreath, now the DHA field secretary, told thousands gathered to celebrate the Americus route in July 1915: "We are going to look to the United States government for aid, for Uncle Sam has been studying the problem."[135] A few months later, in October 1915, the DHA established a three-person committee to oversee progress on federal aid legislation.[136]

As the Dixie Highway took shape, so did a new vision for federal highway aid. Between 1914 and 1916, Congress considered dozens of pieces of highway legislation. As they had for years, these bills varied greatly, from costly proposals for a federal highway system to bills providing modest federal subsidies for county roads. No single bill offered a solution that both Congress and good-roads lobbyists would accept. Conservative congressmen rejected outright the possibility of direct federal aid for interstate highways, but experts and businessmen in the highway lobby had the power to kill overly conservative bills that would have little practical effect on roads.[137]

In 1916 mutual frustration and growing public demand forced Congress to carve a compromise out of two bills, one from each end of the spectrum of proposals that had flooded legislators over the past few years. The first was a more-streamlined version of the bill first proposed four years earlier by Missouri representative Dorsey Shackleford, chairman of the House Committee on Roads. Shackleford's idea for modest federal aid to help counties fund rural post roads was popular among southern congressmen but not the AAA, which was still pushing for a federal highway system, or northern and midwestern congressmen, whose urban districts would receive little or no money under the plan. Their combined opposition had stalled the bill in the Senate after it had passed in the House in 1914.[138] The second bill, proposed by Alabama senator and longtime good-roads booster John Hollis Bankhead in 1915, was crafted primarily by members of the highway lobby and called for a federal highway system. By late 1915, supporters had fused the two very different bills into a kind of bargain whereby millions of dollars in federal aid for rural post roads would be administered by the states. As chairman of the Senate Committee on Post Offices and Post Roads, Bankhead helped to write the final version of the bill and also took charge of shepherding it through Congress.[139]

The Bankhead-Shackleford Bill, as it became known, reflected the sectional divisions that still plagued the Good Roads Movement. On January 25, 1916, the House version of the bill passed with an overwhelming majority of 283 to 81, a margin very similar to that of the first Shackleford vote almost exactly two years earlier. Only northern urban congressmen voted against it. From the Dixie Highway states, the bill faced stiff opposition from several congressmen in Indiana, Illinois, and Ohio, but every single congressman from the other Dixie Highway states voted in favor of the bill. So did every congressman from the neighboring Deep South states.[140] In the Senate, where the rural majority was neutralized, Bankhead had to fight lingering suspicions that federal aid would empower the federal government to make decisions that many people still felt belonged to state and local authorities. During debates over the bill in the spring of 1916, Bankhead repeatedly reassured senators that the bill would not grant any significant new powers to the federal government. "The committee in the framing of this bill," he promised a skeptical senator from Ohio, "kept as far away as possible from Federal supervision, believing . . . it was better to leave the question of construction . . . and location to the State authorities." Yet he admitted that federal oversight was a necessary corollary to federal aid. "[I]f the Federal Government is called upon to put up half the money," he said, "it ought to have something to say about the character of the roads that are to be built." Senators who opposed the bill also argued that any bill that did not provide for the construction of permanent roads was a waste of money. But southern supporters of the bill refused to compromise, knowing full well that their constituents would not be able to match federal funds for expensive paved roads. Senator Hoke Smith of Georgia mocked a proposed amendment by Ohio senator Atlee Pomerene as a stalling technique, saying that by Pomerene's definition of "permanent" roads, "there could be no roads built except those built of Belgian blocks and brick." Growing impatient with the continued debates, Smith interjected: "[I]f any roads proposed to be constructed under this bill will last as long as the debate on it seems destined to last there will be no use splitting hairs as to what the word 'permanent' means." Over weeks of debates in the Senate, it became clear that the overarching concern of most of the bill's opponents was a sectional one. Senator Henry Cabot Lodge of Massachusetts summed up these complaints by arguing that the nation's eight most urban northeastern and midwestern states paid nearly 60 percent of federal tax revenues and should not be called upon to fund roads "in distant parts of the country, under a bill from which we get no real benefit." It was an "extreme injustice," he said, to expect otherwise.[141]

Bankhead's repeated assurances that the bill would serve all good-roads advocates persuaded even his most vocal challengers. After weeks of contentious debate, Bankhead's version of the bill passed unanimously on May 8. Over the next several weeks, a conference committee ironed out the remaining differences between the two versions of the bill, taking care to address lingering concerns over state power and apportionment of federal funds among states so divided in terms of population and geography. President Woodrow Wilson signed the bill into law on July 11, 1916, as the first Federal Aid Road Act.[142] The Good Roads Movement finally had its federal-aid legislation, but it was hard fought, hard won, and born of compromise.

■ ■ ■

Along with the Dixie Highway campaign, strong southern support for good roads shaped the bill that finally made it to Congress in 1916. While so many southern advocates of the Good Roads Movement had long feared that prioritizing long-distance highways would come at the expense of local farm-to-market routes, the bill that was signed into law in 1916 guaranteed the opposite. It was a mixed blessing for Dixie Highway supporters. The bill provided $75 million in federal money over five years, but it left control over construction, improvement, and maintenance of roads and highways to state governments rather than to thousands of county governments. Every state that did not already have a state highway department would have to establish one in order to receive federal funds. In order to assuage the various rural and urban, northern and southern proponents of the bill, appropriations would be based on population, mileage, and land area. And the law specifically limited federal aid to rural post roads. Cross-country automobile highways like the Dixie would benefit only when and where sections of the highway through individual states could be used as local postal routes as well. Most significantly, however, the law did not require federal-aid roads to be linked to one another, further limiting the bill's impact on interstate projects like the Dixie Highway.[143]

The Federal Aid Road Act was an unprecedented demonstration of federal support for road building. It signaled the U.S. government's recognition of the Good Roads Movement and, despite some serious limitations, projects like the Dixie Highway, whose supporters included a number of the same automakers, lobbyists, state highway officials, and ordinary citizens who backed the new highway bill. But the new legislation also embodied the reservations that many people still had about federal intervention. By funding only post roads, the bill preserved local governments' responsibility for the vast majority of the nation's roads. And by not requiring federal-aid roads to be interconnected, it made the work of groups like the Dixie Highway Association all the more difficult. The full vision of a federally funded interstate highway system was, in 1916, too radical and too foreign a project to garner congressional support.

All of this would change when the outbreak of World War I and the demands of wartime preparedness made the prospect of greater state and federal control over interstate highways appear more practical—a matter of national interest rather than simply the will of a group of citizens in the Good Roads Movement. The DHA pounced on the opportunity opened up by the wartime emergency. World War I created a platform from which the DHA could launch a new and invigorated campaign for federally funded interstate highway construction.

· 3 ·

Roads at War, 1917–1919

On October 30, 1917, nearly seven months after the United States entered the war against Germany, the Dixie Highway Association staged a test run of military supplies along its eastern and western divisions between Fort McPherson, outside of Atlanta, and Fort Oglethorpe, near Chattanooga, where three camps of U.S. Army recruits learned trench-warfare tactics under a young officer named Dwight D. Eisenhower. The dirt roads they utilized bore little resemblance to the interstate highways that Eisenhower, as president, would oversee some forty years later, but after two years of repairs and improvements by local road authorities, they were substantially better than the wagon trails they once had been. With clear, dry weather to their advantage, the military supply trucks arrived in record time. The DHA estimated that the five two-ton trucks carrying materials saved $1.62 per ton and twelve to thirty-six hours over the same trip by railroad. The truck carrying the soldiers and their gear saved at least three hours and $2.89 per man.[1] In the midst of the nation's largest and costliest war to date, time and money were resources that the U.S. government could scarcely afford to waste.

Although the DHA reveled in the results of the army truck tests, it also used them to raise the stakes of the Good Roads Movement by calling for the construction of permanent long-distance highways to ensure that the nation could meet wartime demands in any weather. "If we are going to have our country in a true state of preparedness and efficiency," one DHA representative wrote, "we must not be so situated that we have to hold our breath lest our roads be temporarily incapacitated every time it rains." The truck tests were part of a larger strategy by which the DHA launched a comprehensive wartime campaign to demonstrate that modern interstate highways were cheaper and more reliable than railroads for transporting both people and goods.[2]

The First World War provided a valuable opportunity for the DHA and road advocates to carve out a more-profitable relationship between the Good Roads Movement and the federal government. By the time the United States entered the war on April 6, 1917, every state had established a state highway department in accordance with the guidelines of the Federal Aid Road Act of 1916.[3] But in the South, these fledgling agencies remained weak and underfunded for years to come. More than a year after Congress passed federal funding for roads, county governments still controlled virtually all of the work in the South, including work on the Dixie Highway. They were eligible for federal aid for portions of the highway only when they overlapped with rural post roads. If they could not afford to pay their half of the construction costs, they received nothing. Even when they did qualify for federal aid, their mostly dirt roads could not stand up to the demands of heavy traffic, and they were prohibited from spending federal money on maintenance.[4] Travelers found themselves at the mercy of local roads and the weather, just as they always had been.

Throughout the war, in hopes of securing federal aid, the DHA recast interstate highways as military necessities and not just tourist arteries or farm-to-market routes. By defining highways as federal investments, the DHA suggested that interstate highways were vital to both the national economy and national defense. Their strategy had mixed results. Congress did not significantly alter federal aid for highways during the war, but the war forever changed how politicians and citizens thought about highways. In 1919, just after the war ended, the U.S. Army sponsored its own military truck test. This test, a transcontinental convoy of eighty-one trucks and 300 soldiers, only crossed over the Dixie Highway, but it nevertheless validated the DHA's test of two years before. One of the participants was Dwight D. Eisenhower, who later said the arduous trip had convinced him of the need for long-distance highways. Indeed, the convoy was the federal government's way of showing its commitment to the Good Roads Movement and, more specifically, to the benefits of trucks and highways to move men and equipment.[5] It could not have been a more fitting conclusion to the DHA's wartime campaign promoting those very same things. The war forced everyone, including the federal government, to rethink the scope and purpose of federal aid as it became more and more clear that the 1916 bill was far too limited to meet the demands of either the wartime mobilization or the increasing commercial traffic. When the war ended, the nation was poised to begin building an interstate highway system, with the Dixie Highway as its backbone.

A Wartime Window of Opportunity

In 1915 a *New York Times* writer expressed the hopes of many when he wrote that the Dixie Highway "will serve as a new bond of sympathy between the States and a new means of industrial development . . . [and] should be built to endure, like the old Roman roads still traveled in Central Europe, and symbolic of the united strength of a great nation."[6] The DHA followed suit by circulating a poem about the famous network of stone roads that played a critical role in building and sustaining the mighty Roman Empire for centuries:

> When Caesar took a westward ride
> And grabbed the Gauls for Rome,
> What was the first thing that he did
> To make them feel at home?
> Did he increase the people's loads,
> And liberty forbid?
> No; he dug in and built good roads—
> That's what old Caesar did [. . .]
> He built good roads from hill to hill,
> good roads from vale to vale;
> He ran a good roads movement
> Till Rome got all the kale;
> He told the folks to buy at home,
> Build roads their ruts to rid,
> Until all roads led up to Rome—
> That's what old Caesar did.
>
> If any town would make itself
> The center of the map,
> Where folks will come and settle down
> And live in plenty's lap;

If any town its own abodes
Of poverty would rid,
Let it go out and build good roads—
Just like old Caesar did.[7]

Notwithstanding the eventual fate of the Roman Empire, the poet was right about the historical relationship between military might, economic strength, and good transportation routes. Mindful of the war in Europe even before U.S. involvement, Dixie Highway boosters invoked these historical linkages to promote federal aid for modern military highways.

For many reasons, the moment seemed ripe for new attention to highway development. Perhaps the most significant reason was the growing concern over the inability of existing rail lines to handle the demands of wartime mobility. Despite the growing popularity of automobiles and marked trails in the prewar years, powerful railroad companies still dominated interstate trade and travel. Rail-line mileage grew apace with expanding populations and westward-moving markets, peaking at 254,037 miles in 1916. But the numbers obscured the railroads' financial difficulties in the years leading up to the war. The Hepburn Act of 1906, a Progressive-backed bill intended to reign in railroad abuses, effectively prevented railroad rate increases, keeping revenues static during a period when prices for steel, fuel, and labor rose steadily. When war broke out in Europe in the summer of 1914, the railroads proved ill equipped to handle the sudden spike in shipments, despite significant mileage increases. European demand for U.S. food and munitions created a 43 percent surge in railroad freight. Both the Interstate Commerce Commission and the railroads attempted to address the new wartime challenges with advisory boards and internal regulations, but to no avail. Too few railcars, aging terminals, and uncoordinated schedules among the nation's many railroad lines created massive cargo bottlenecks for both commercial and military traffic.[8]

Things reached a crisis point when Germany resumed unrestricted submarine warfare in February 1917, targeting American vessels and all but stopping transatlantic shipments. Railcars packed with supplies backed up around eastern ports. After the United States entered the war, the army had a difficult time getting its own troops and supplies to the East Coast for deployment. By the fall of 1917, 180,000 freight cars sat waiting to be shipped overseas, creating a shortage of some 158,000 railcars nationwide. With early and severe winter weather beginning to slow shipments even further, everyone realized that extreme measures were necessary to prevent a complete shutdown of the nation's only supply lines.[9]

The railroad crisis led to dramatic and highly controversial transformations in the relationship between the federal government and transportation networks. On December 28, 1917, President Woodrow Wilson ordered that the nation's railroads pass under federal control for the first time in history. The railroads were not nationalized, but a temporary wartime Railroad Administration under the guidance of Treasury Secretary William G. McAdoo invested in 100,000 new freight cars and nearly 2,000 locomotives and centralized control over the routing and distribution of railcars. Eventually, the government also suspended all unnecessary civilian traffic by limiting passenger service, a wildly unpopular move. The government justified this drastic wartime measure because there were no alternatives to long-distance overland shipping. During the war, privately owned railroads became vital in the service of public interests.[10]

Leaders of the Good Roads Movement viewed the wartime railroad crisis and the subsequent government takeover of rail as their chance to argue for greater federal investment in transportation alternatives. However temporary such emergency measures may have been, they realized, the wartime transportation crisis could permanently alter the extent of federal involvement in highway construction. For Dixie Highway counties still struggling to meet the demands of modern traffic under the limitations of the 1916 Federal Aid Act, the wartime emergency provided an unparalleled opportunity to publicize the structural limitations of a railroad monopoly over long-distance transportation. It was in this national conversation that the DHA found new opportunities to lobby for the advantages of investment in highways and automobiles as necessary and potentially superior alternatives to rail.

But DHA officials had their work cut out for them. For all the shortcomings of rail, roads—not to mention the resources of local road commissions—were in worse shape. Despite progress along the Dixie Highway, the roads that comprised it and dozens of branches and feeder roads remained far from complete when the United States entered the war in 1917. The summer before, the DHA's board of directors stripped one route in middle Georgia of its official designation after the majority of counties along the route failed to do the work they had promised. The directors rerouted the highway through neighboring counties, and they threatened to similarly punish another county if it did not complete critical repair work within ninety days.[11] These failures reflected the shortcomings of local funding for infrastructure projects, as well as the 1916 Federal Aid Road Act.

Faced with a national crisis, the DHA began to transform its approach to lobbying for federal highway aid. By presenting roads as wartime necessities, the DHA hoped to secure significantly more federal support for long-distance highways. Although such an approach challenged deep-rooted, ongoing opposition to federal control over roads, northerners and southerners alike recognized that the nation's local road systems could not meet military demands. The DHA launched a massive propaganda campaign to reinvent the Dixie Highway as the foundation of a future system of national military highways.

The Propaganda War

The Dixie Highway Association began to reshape its strategy as early as the summer of 1916, when the directors turned their occasional eponymous publication, *Dixie Highway*, into a glossy, colorful monthly magazine. Overnight, the DHA stopped promoting the Dixie as merely a tourist route and market road and began celebrating it as an investment in national defense. In April 1917, the same month President Wilson asked Congress for a declaration of war against Germany, *Dixie Highway*'s lead article proclaimed: "Good Roads a War Time Necessity." Raising the possibility of a German invasion along the East Coast, the author, Colonel J. P. Fyffe, argued that the United States would need to have reserve soldiers stationed inland, near good north-south roads. A map in the center of the article labeled the Dixie Highway "The Road behind the Firing Line," a route ideally situated to "serve a section manufacturing and producing all necessary munitions and army supplies" and also nestled safely behind the Appalachian Mountain range, shielded from coastal attack. Dixie Highway crossroads such as Atlanta and Chattanooga were ideal locations to station soldiers and supplies, Fyffe concluded, because they were far enough from the front line to be safe but close enough to get men and supplies there quickly.[12]

This opening salvo in the DHA propaganda campaign exposed and exploited the railroads' shortcomings in order to assert the fundamental superiority of long-distance highways. Although railroad lines were already in place to serve in case of an East Coast attack, Fyffe cautioned that the current "heavy guard" stationed at the country's railroad bridges was "but a reminder that the railroads might easily be destroyed as a means of transportation, in which event the army would be compelled to rely upon motor trucks, which in turn depend upon the condition of the roads." He directed his message squarely at southerners, whose roads were least prepared for such an emergency. "If the people of this section are anxious to contribute something to the preparedness for the nation's defense," Fyffe chided, "they can do it in no better way than by furnishing a highway which will put this natural strategic point in communication with the factories of the North and the graineries [sic] of the West."[13]

Crucial to the DHA's message was the warning that public faith in rail provided a false sense of security. In July, just three months after the United States entered the war and while the federal government struggled to manage the railroad logjam, the DHA's business manager, V. D. L. Robinson, jumped at the opportunity to point out the deficiencies of rail travel. "With a nation so accustomed to look upon the railroad as the most important of all means of transportation," he said, "the average citizen is inclined to feel that the President, and the national defense committees, by taking drastic steps to eliminate delays in railway traffic has the solution of the problem well in hand and that everything possible is being done." But Robinson reminded readers that such a view was shortsighted and careless. He estimated that "the highways of the Nation represent eighty-five percent of the transportation facilities while the railways represent fifteen percent," and the former were in even worse shape than the latter. A person needed "only to step out of his front gate to be face to face with the big end of the most important problem connected with the war."[14] This same sense of urgency pervaded the wartime issues of the DHA's magazine. Cover art declared the Dixie Highway "A National Necessity," while feature stories issued warnings that rang out like alarm bells. "The need for action, and immediate action," one writer cautioned, "has never been greater."[15]

The magazine often argued that railroads could barely handle existing commercial traffic, much less simultaneously manage regular shipments and military needs at the same time. As one writer pointed out, in 1917 only four lines served the entire Upper South. Were any of these lines to become disabled, military camps throughout the rest of the South would be cut off to supplies from outside the region. In September 1917, just five months after the nation entered the war, the magazine reported that some of the southern railroads were forced to stop commercial shipments altogether in order to help the army move a mere 5 percent of new draftees to camps. "With the actual transportation of the main body of troops," an editorial warned, "the situation will become most acute." Northern shippers who needed to get materials to the South tried to "dodge" the congestion by using old-fashioned express companies, which were ill equipped to keep up with long-distance commercial-shipping demands. "With this knowledge of traffic conditions," the writer concluded, "one can readily see that the most patriotic and valuable service which good roads enthusiasts can render is the properly directed effort to build and maintain the main highways."[16]

Properly directed roadwork required significantly more federal aid than the Federal Aid Road Act of 1916 provided. Dixie Highway boosters hoped their wartime propaganda would give interstate highways the same kind of significance at the federal level as the railroads. A full-page cartoon in the October 1917 issue of *Dixie Highway* made clear that the DHA wanted the federal government to recognize the urgency of federal aid for long-distance highways. In the foreground of the cartoon, a cargo truck carrying military supplies drove along a smooth Dixie Highway, while in the background, motionless railroad cars overflowed with "war freight congestion." Looking over the scene, Uncle Sam declared: "The 'Dixie' is my right hand man."[17]

The DHA believed that securing federal aid for interstate highways would be easier if they could sell the federal government on the idea that the Dixie Highway was first and foremost a military route, even if it also served other purposes. The DHA thus sometimes downplayed fragile road conditions and instead portrayed the highway as an already-integral part of the military's transportation system. "If an aeroplane squadron should be ordered to take observations on the territory between Chicago, Illinois, and Jacksonville, Florida," they bragged, "they could make an amazing report on the cars, motor trucks, motorcycles, etc. that are speeding daily between the eight great cantonment sites situated on the Dixie Highway as they journey about the big business of getting Uncle Sam ready to go to war." The DHA declared that those eight camps made it certain that the Dixie Highway "is destined to become the military route of the East Coast."[18]

Many other groups joined the DHA in using the war to promote bet-

Cartoon, *Dixie Highway*, October 1917.

ter funding for national highways. On the eve of the war, supporters of the Dixie Highway, the Lincoln Highway, and several other marked trails (such as the planned Jefferson Highway between New Orleans and Winnipeg, Manitoba) banded together to form the National Highways Association (NHA). Members declared that there could be "no real preparedness for war, for defense, or for peace without national highways and good roads everywhere."[19] More-established organizations supported the DHA's efforts to lobby for wartime highway spending as well. Two months after the United States entered the war, the American Automobile Association called for the construction of a "marginal" military border road that encircled the entire country. Since the 1916 Federal Aid Road Act did not provide funding for such a highway, the AAA "urge[d] upon Congress the need of legislation to provide for a system of military marginal roads to be constructed and maintained at national expense."[20] The AAA even suggested that the war had soothed lingering tensions highlighted by the Dixie Highway routing contest and the fight over the 1916 federal aid bill. AAA president George C. Diehl argued near the end of the war that "war conditions have emphasized national transportation needs to the extent of obliterating State lines and sectional boundaries." Therefore it was "logical," he said, that the 1916 federal aid law be amended to give federal officials more control over roadwork. "The great possibilities of main artery highways," he believed, "are becoming apparent to every citizen who wants to see the war brought to its earliest conclusion."[21]

While interstate routes transcended regional boundaries, the South's distinct advantages as a center for wartime mobilization made it particularly good as a focus for the wartime highway campaign. Climate was a major benefit in the DHA's arguments for federal funding along the Dixie Highway. *Dixie Highway* called the South "the great army field for the winter" because the region's moderate weather allowed for year-round training of new recruits. Six of the eight army camps along the Dixie were in Louisville, Nashville, Chattanooga, Atlanta, Macon, and Jacksonville. As mobilization increased, the army built even more camps in the South. By October 1917, there were five camps in Georgia alone; all were along the Dixie Highway, including Fort Screven in Savannah, which held essential coastal artillery. These camps were "hustling, busy, energetic little cities" that increased the state's population as well as the demand for food, clothing, shelter, and consumer goods.[22] Transportation demands in Georgia rose to an all-time high during the war, making traffic along the Dixie's dirt roads heavier than ever and showcasing the shortcomings of a highway that was so important to the war effort but still far from complete.

The military itself could scarcely ignore the importance of maintaining the country's most critical routes. Whenever possible, the DHA incorporated statements of support from military personnel to bolster its good-roads propaganda. Major General Henry T. Allen proclaimed the Dixie Highway to be "a military asset," but he implied that it needed work before the army could tap its full potential. "With its numerous manufacturing cities and its vast supply centers," Allen predicted, "the Dixie Highway would offer most advantageous sites for supply bases and would constitute a highly important base line, sufficiently far from the coast to be reasonably safe from air crafts, yet sufficiently close for many war requirements." *Dixie Highway* featured Allen's remarks alongside a cover image showing military trucks on the highway whizzing past factory smokestacks.[23] Military personnel frequently expressed support for better roads and highways in the national media as well. Major General Leonard Wood told one good-roads booster writing for the *New York Times* that a "systematic, well-planned network of roads" was not only a good investment for commercial purposes but also "of tremendous value to our military forces." He added that "through routes connecting centres of production of population aid in the prosecution of the war" by making it easier to move men and supplies, and he urged the government to consider building modern paved highways along the Eastern Seaboard.[24] Though rare, the military sometimes even put its money where its mouth was. In September 1917 the War Department sent railroad cars and other equipment to workers building a critical eleven-mile stretch of the Dixie Highway through Boone County, Kentucky. The strategic road linked Fort Thomas and a new army camp in Louisville a hundred miles to the southwest.[25]

Interstate highway proponents had little trouble explaining their shift in focus from tourist highways to military routes. While the DHA admitted that the main impetus for initiating the Dixie Highway had been tourism, they argued that this had been "augmented" by the nation's wartime needs for the Dixie as "an auxiliary, and if need be, a substitute for the railroads" in the mobilization of goods for the war effort. And although the directors were well aware of the difficulties of doing more roadwork without more federal aid, the DHA declared that the counties along the highway had a "solemn obligation" to speed up their work "as an act of patriotic preparedness."[26] Outside the South, Lincoln Highway boosters followed suit and emphasized military preparedness over tourism. "War conditions," one observer said, made people more aware of "the value of the Lincoln Highway as an auxiliary to the railroads" and not just a tourist highway. Supporters of both highways exaggerated their own preparedness for heavy military traffic, but they did not overestimate the value of patriotic fervor to the Good Roads Movement.[27]

A year into the real war in Europe, the DHA's propaganda war culminated in a unanimous resolution urging Congress and the president to create a "centralized Federal Authority" to develop and direct a new highway policy that better met wartime demands and to coordinate transportation along the nation's highways, railroads, and waterways.[28] While it echoed the highway lobby's decade-old plan for a federal highway commission, the DHA proposal went even further by acknowledging that highways were but one part of a massive transportation network that could not be managed piecemeal. The resolution carried little weight coming from an unofficial—and obviously self-interested—organization like the DHA, but it nevertheless evoked widespread concerns about how wartime traffic had utterly overwhelmed the nation's transportation capabilities.

The proposal also reflected worries closer to home. Despite the sincere desire to promote interstate highway construction nationwide, the DHA's propaganda campaign was motivated primarily by the desire to complete the Dixie Highway. The war had made that all the more difficult, as counties and the new state highway departments grappled with greater traffic demands and stagnant local budgets. Even as they promised to ask Congress for a federal highway board, the DHA was pleading with William McAdoo of the Railroad Administration for permission to use railcars to carry construction materials to dozens of counties along the Dixie Highway. Without the railroads, work on the highway was at a standstill. The DHA had little choice but to give up on some communities altogether. In May 1918 the directors revoked another designation along the highway through Kentucky and eastern Tennessee when several counties there failed to keep up with their share of the work.[29]

As the DHA struggled to transform local roads into something befitting a national military highway, it collected new partners in the crusade for federal highway funding. Their joint mission was familiar to longtime supporters of the Good Roads Movement. But in the context of war, it took on new meaning and also afforded new opportunities to pursue the longstanding goals of good-roads supporters for an integrated system of federally funded, federally managed interstate highways.

The Highway Business

In September 1917, the U.S. Chamber of Commerce held a "War Convention" in Atlantic City, New Jersey. Despite pledging "vigorous support of the war at any cost," the chamber was transparent about its economic concerns. "War disrupts business," one merchant group said, and this made necessary "continual adjustments" in business practices in order to survive the wartime emergency.[30] Businessmen accepted government intervention in the economy during the war, even to the point of fixing prices and raising taxes. To them, these measures were necessary to ensure continued profits during the economic upheaval created by the war. But they also wanted the government to give them something in return. Delegates to the convention devoted hours to discussing the role that transportation played in the national wartime economy, and they concluded that improving the nation's roadways was "important and essential as a war measure and should be urged in every possible way."[31]

For the business community, the wartime campaign to improve roads was as much about the national economy as it was about national security. Federal investment in long-distance highways would not only help a nation at war; it also would help businesses to profit during the war by providing soldiers and citizens with the goods they needed. Farmers and small-town businessmen alike had complained about the high cost of the "mud tax" for years, but the war gave new life to the economic argument for roads. Just weeks after the United States entered the war, DHA business manager V. D. L. Robinson estimated that poor roads cost the nation's businesses "the appalling sum of *more than three hundred millions of dollars a year.*" He asserted that "the reduction of waste is more important than ever before with the nation at war." Robinson did not advocate immediate improvement of all roads but rather the concentration of attention on main highways "serving the largest percentage of traffic and leading from one center of population and production to another." The Dixie Highway, he continued, "extending from the Great Lakes to the Gulf and connecting the leading centers of population and production of the Central States as well as the majority of training camps, is one of the most important highways from the standpoint of military necessity in the United States."[32]

The business community, long aware of the structural inadequacy of roads and the financial limitations of rail, capitalized on the wartime campaign to advance their own interests. Some manufacturers began experimenting with trucks on freight routes between nearby cities as alternatives to railroads, and, where it was possible, farmers of perishable food crops began using trucks for shipments of milk and produce. In Louisville, Henry C. Kelting expanded his small intercity trucking business during the war in order to haul foodstuffs to three nearby army bases, most likely using the Dixie Highway.[33] Businesses in the West also turned to trucking companies both to alleviate freight congestion and to replace railroads in small western communities where there was limited rail service. The Tonopah Trucking Company in Reno, Nevada, noted: "The business of our clients would cease to be if they had to rely solely on railroads." A midwestern produce salesman also reported using trucks successfully in small towns that were not on railroad lines. Despite such promise, most shippers complained about the limits of existing roads.

Even in the Northeast, home to some of the nation's best roads, shippers lamented the inefficiency of freight transport along rough highways. The Oriental Silk Printing Company in New York City, for example, used trucks to haul silk to and from nearby Paterson, New Jersey. "Bad roads [are] interfering," a company official admitted, but they could not rely "solely on railroads" to do business.[34]

Companies using trucks to replace or supplement railroads were only too aware of the constraints of local roads. Alexander Morton, who hauled lumber for wartime industries in Virginia, reported in December 1917 that within the month, "the roads will be so bad in that locality that I will have to abandon the use of trucks until spring, and rely on teams to do my hauling." Even in Chicago, a cement company official complained that poor highways restricted the company's use of trucks to local deliveries. Even where trucks proved useful early on, it quickly became clear that heavy trucks could not drive repeatedly over most roads. An official for the United Gas Company in Philadelphia hauled supplies to New York on roads that deteriorated under the increased wartime traffic.[35] An engineer for the Office of Public Roads complained that within just a few weeks, hundreds of miles of vital roadways "failed under the heavy motor traffic." He described how "almost overnight, an excellent surface might become impassable" under the stress of wartime traffic.[36] Yet there was little his agency could do about it. In July 1918 a *Dixie Highway* cover cartoon bluntly assessed the predicament faced by shippers and sympathetic public officials alike with a rhetorical question: "We have the trucks, have we the roads?"[37]

Within the business community, automobile men were among the first to take advantage of these critiques. State highway departments in the South were newer and thus even more ill equipped to handle the stress of wartime traffic than their northern and midwestern counterparts. Seeing an opportunity to help them and also draw attention to the lack of funding, auto manufacturers began to fill this critical gap in state support for roadwork by planning, surveying, and helping to fund important sections of the Dixie Highway. And like the DHA, these men explicitly linked their work to the war effort. In January 1918 the wealthy members of the Chattanooga Automobile Club put their considerable resources to work "to speed up the opening of highway communication between the North and the South" by helping to complete the western division of the Dixie Highway between Nashville and Chattanooga. The club raised money for this link in the highway through a membership campaign, which doubled its numbers to 1,000. To mark the occasion, members voted unanimously to give another $1,000 to the Dixie Highway Association.[38]

The Chattanooga Auto Club's private investment in public highways did not end with its membership fundraising campaign. When DHA president Michael Allison told club members where roadwork along the Nashville-Chattanooga link lagged, one member offered to donate $750 if the club could match it. It did, and members began to draw maps, organize data, and plan scouting trips to find out precisely what it would take to make the western division between the two cities passable. Aware that they could not singlehandedly master such a large-scale project, members announced their intention to take all the information they gathered to the next meeting of the Tennessee State Highway Department in Nashville. They succeeded, and later that month, the highway department agreed to appropriate $300,000 of the state's share of federal aid for this stretch and another on the eastern division route.[39] Instances like these were rare but remarkable examples of how the DHA and its powerful supporters could harness state and federal resources when they needed to.

The DHA may have shifted its focus to military preparedness, but tourism remained an important incentive in the campaign for federal highway aid. Although the Nashville-Chattanooga road linked military bases in the Midwest to vast numbers of army camps in the South, it was equally important as a tourist and trade route from the West and Midwest to the Deep South. Every issue of *Dixie Highway* included a section of "Touring Queries" in which primarily nonsouthern tourists sought advice about the condition of various roads in the Dixie network. Magazine writers went out of their way to advertise certain cities as ideal tourist destinations. Milledgeville, Georgia, the quaint former state capital, was a popular stopping point for tourists along the Old Capital Route to Savannah. Thomasville, Georgia, where old plantation homes had been converted into hotels many years earlier, remained a popular resort for northern tourists. On the Waycross route through southeast Georgia, the small town of Fitzgerald was billed as the "Yankee Town of the South" thanks to an industrious northern developer who years earlier had established a retirement community for Union veterans of the Civil War.[40] Auto club members like those in Chattanooga were more interested in tourism than most because many of them ran auto industry–related companies, and most were involved one way or another in local Chattanooga businesses.

Their backgrounds, every bit as much as *Dixie Highway* write-ups with titles such as "The Motor Tourist and the Dollar Sign," left little doubt about what motivated many good-roads advocates both before and during World War I.[41]

Eager to appear patriotic rather than self-seeking, members of the Chattanooga Automobile Club denied having any financial interest in improving the Nashville-Chattanooga link, and the DHA backed them up. "A striking feature" of the club's meetings, the DHA boasted, was that members gave "little consideration" to the material benefits of tourism along the Dixie Highway. Instead, it was "sufficient for them to know that this road is needed to help the Nation in winning the war." With their scouting tour coming up, club members declared magnanimously that "if the soldiers could work in the trenches during the winter weather," they themselves "should be willing at least to endure the few days of discomfort necessary to make a personal inspection" of rough road conditions in order to "more effectively carry on the [wartime] campaign." However insincere their claims may have been, the active roles that both the DHA and private business organizations like the Chattanooga Automobile Club played in promoting tourism during the war was valuable to the wartime campaign, in addition to being a testament to the shortcomings of local, state, and federal highway agencies in a time of national emergency.[42]

Other longtime members of the highway lobby also hoped to profit from the wartime emergency, and the DHA backed them up. The cement industry was one of the largest and most vocal. Since the routing campaign two years earlier, Dixie Highway advocates had been urging investments in hard-surfaced roads, but increased traffic from tourists and the military made it all the more necessary. The closest material to modern-day asphalt was a combination of broken stone and tar named after John McAdam, the Scottish engineer who had invented it a century earlier. However, experts were divided over whether or not macadam roads, as they became known, were durable and smooth enough for heavy automobile traffic. The only viable alternative was cement, which before 1917 had been used only sparingly in and around cities due to the cost.[43] In April 1917 the "steady flow" of inquiries about the best routes between midwestern cities and Florida suggested to the DHA that rural traffic was increasing rapidly, as well, and that "durable roads" were becoming "a necessity from the standpoint of economy." The DHA argued that cement roads were the highways of the future and predicted that millions in maintenance would be "utterly wasted" if Americans continued to invest in "roads that wear out as fast as they are built." With passengers traveling nearly 10 million more miles a year by car than by railroad by the time the United States entered the war in 1917, the need for modern roads was more pressing than ever before.[44]

Wartime propaganda was aimed at voters as well as consumers. *Dixie Highway* frequently carried advertisements from cement companies urging the public to support the construction of "permanent roads." Often the ads placed the companies squarely in the midst of the debate over local versus state and federal funding. Knowing that cement roads cost much more to build than did dirt or gravel roads, the Universal Portland Cement Company, based in Chicago, ran an ad praising the way one Indiana county raised the money to hard-surface its roads. "Vermillion County Invests $19 Per Capita in Permanent Roads," the ad title read. "Bond Issue Makes New Highways Immediately Available."[45] Though hard-surfacing was less common in the South, southern companies also encouraged cement-highway construction financed through bonds. The Dixie Portland Cement Company, headquartered in Chattanooga and owned by DHA founder Richard Hardy, published advertisements encouraging citizens to use the ballot box to demand better roads. "Voters Should See to It Now," one ad read, that their roads be built "properly" using funds provided by bonds. "If there is a good roads bond issue up in your County or State," the company begged, "vote for it; if there is not, ask your legislators and road authorities to submit a bond issue to the people."[46]

If established companies hoped to benefit from the construction of a modern, surfaced interstate highway, smaller and newer businesses depended upon the success of the DHA's wartime campaign for their very survival. Auto tourism was still in its infancy in the late 1910s, but already the increase in traffic had led to the construction of new filling stations, auto garages, roadside stands, and small mom-and-pop stores. Simple, no-frills campsites called auto camps (or sometimes "gypsy camps") sprang up almost overnight on major tourist routes like the Dixie Highway. These spots were popular with tourists short on money and also among those who wanted to experience the freedom of stopping anywhere they pleased for the night. But mostly they were a necessity: bad roads and still-unreliable cars made it difficult for tourists to count on making it to the next city by dark. To satisfy wealthier motorists with

A CONCRETE ROAD UNDER CONSTRUCTION

CONCRETE ROAD — LEE COUNTY, MISSISSIPPI

Good Roads Will Relieve a Food Crisis

We Must Do More Than Raise Big Crops
—*We Must Get Them to Market*

A system of roads strong enough to bear a big volume of traffic is positive insurance against extreme food shortage.

However short the crop, food of some kind could then be promptly moved, without danger of being blocked or spoiled in transit; and when it arrived in town, could be immediately delivered. There is food enough in the country if it can be freely moved from distant as well as nearby farms. A large tonnage goes to waste or is marooned by bad roads. The farmer loses his market and the consumer pays an exorbitant price.

Short Haul Trucking on Good Roads

will supplement long haul railroad traffic. Motor trucks are cheaper, quicker and more efficient for short distance work. They travel under their own power, independent of schedules, tracks, yards or locomotives. But they require roads as durable as concrete. Nothing less rigid will stand the pounding, thrusting and shearing effects of power-driven truck wheels. Scattered stretches of good roads are insufficient.

Voters Should See to It NOW
That a system of highways be properly laid out in their County or State.

That money enough be raised by a bond issue to build them now, and build them right.

That nothing less durable than concrete be adopted for all motor traveled roads. It has the strength and durability which makes it so universally used in great engineering works.

If there is a good roads bond issue up in your County or State, vote for it; if there is not, ask your legislators and road authorities to submit a bond issue to the people. The tax levy to pay it off is so light it won't be felt — probably 5c to 10c per acre of farm land and 50c to $1 per town lot annually for a period of 20 years, after which the debt is wiped out and the roads continue to afford good service.

Write Us For Copy of "CONCRETE HIGHWAY"

DIXIE PORTLAND CEMENT COMPANY
JAMES BUILDING   CHATTANOOGA, TENNESSEE

Advertisement, *Dixie Highway*, April 1917.

more-refined tastes, hotels and motels, initially rare except in railroad towns where travelers changing trains might need an overnight stay, popped up at more-frequent intervals along the road.[47] One of these fine hotels was the Hotel Patten in Chattanooga. Built by Chattanooga Medicine Company owner and DHA founder James A. Patten in 1908, the hotel served as the headquarters of the DHA. Frequent advertisements for the Patten and other luxury Dixie Highway hotels filled the back pages of *Dixie Highway*.[48] In a region where agriculture was becoming less and less profitable, the tourist business even kept some farm families afloat. They took in boarders and sold produce at roadside stands to supplement the family income.[49] Sometimes, tourist traffic gave rise to lucrative new trades. In the textile-mill town of Dalton, Georgia, a cottage industry for hand-tufted chenille bedspreads arose around the turn of the century, but it took off after the Dixie Highway came through. Women began selling the colorful spreads at roadside stands that lined the highway for miles. By the 1930s, the homegrown industry had attracted northern capital and evolved into a thriving textile business employing some 7,000 local workers.[50]

Along with the automobile, the war helped to shape a new and sophisticated domestic tourism industry. The "See America First" campaign had begun during the Gilded Age and prospered with increased federal funding for national parks under President Theodore Roosevelt, and the advent of marked trails in the 1910s energized it. The war, however, made it a national pastime, especially among wealthy and middle-class Americans. The war interrupted travel abroad starting in 1914, so tourists treated traveling and spending money at home almost as a "patriotic duty," as one historian has argued, and "a ritual of American citizenship."[51]

While national parks out west drew thousands of tourists during the war, the South became a very popular destination as well. The Dixie Highway Association estimated that 27,000 motorists in nearly 7,000 cars spent $2,760,000 along the Dixie Highway during the winter season of 1916–17 alone, though most were bound for Florida.[52] In order to market the South to even more northern tourists, Dixie Highway boosters and businesses had to present the region as an attractive destination in and of itself, not just an obstacle separating Chicago snowbirds from the warm Florida sunshine. The very name they had chosen made the highway sound like a road to the past as much as a road to a particular place, and businesses along the Dixie Highway made the most of it by selling tour-

Hotel Patten advertisement, *Dixie Highway*, April 1918, 12; Hotel Savannah and Hotel Seminole advertisements, *Dixie Highway*, March 1917, 12.

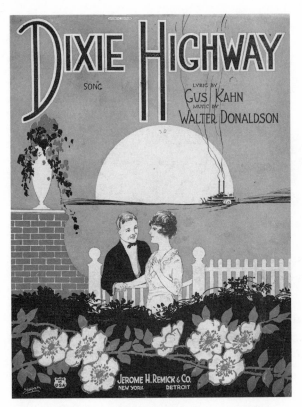

Cover of sheet music for "Dixie Highway," published in 1922.
(Courtesy of Chattanooga History Center)

ists on the image of the South as an exciting and exotic locale, a romantic ideal to be explored and exploited. In Illinois and Indiana, "Dixie" gas stations, restaurants, and hotels conjured up images of an unfamiliar but pleasant destination.[53] Many of the appeals to tourists that had fueled the routing contest two years earlier were beginning to bear fruit by 1917 as well. The DHA encouraged northern motorists to tour south Georgia's peach and pecan orchards and rhapsodized about driving country roads lined with the region's "fragrant pines."[54] Civil War tourism also drew many northern tourists. While the industry was most profitable in cities like Richmond, the Dixie Highway lured tourists to sites in Georgia as well.[55] In time, composers even began to write songs about the Dixie Highway, while picture postcards depicted pleasant drives through beautiful scenery.[56]

Postcard, ca. 1920s. (Author's personal collection)

The DHA published regular updates on road conditions in the monthly magazine, but many tourists relied on published guides to get them where they wanted to go. Before the war, auto-related businesses sponsored some of the earliest road maps and automobile guides to point tourists in the direction of the nearest hotels, garages, restaurants, and interesting landmarks, as well as to guide them along the mostly unmarked roads and marked trails. The most popular of these was the *Official Automobile Blue Book*, published annually from 1901 to 1929. These small booklets featured hand-drawn maps, mileage estimates, and lengthy descriptions of roads and landmarks to guide tourists as they hopscotched from town to town along bumpy dirt roads. Although they offered simple, perfunctory instructions, auto guides also warned motorists about steep grades, misleading signs, and other obstacles. But early guides were often expensive and inaccurate, so tourists were at the mercy of the publishers.[57]

As one historian has argued, the significance of World War I as a stimulus to the creation of modern, reliable road maps is difficult to overemphasize.[58] The escalation of both tourist traffic and military traffic during the war increased demand for more-accurate maps and guides, so mapmakers and auto industrialists began publishing a new kind of automobile road map. Groups like the Lincoln Highway Association and the Dixie Highway Association commissioned maps of their own routes, but more comprehensive maps of interlinking marked trails were more common and more popular.[59] The very first Rand McNally map of marked trails was published in 1917. In place of the lengthy descriptions that characterized auto road guides, these maps used standardized symbols to indicate the locations of hotels, garages, and different types of roads and highways. Instead of close-up views of local sections of roads, these maps showed regional or even national networks of roads, rough and incomplete as they were. Road signs were rare, but as wartime traffic picked up, Rand McNally and other companies also began to assist in marking the many marked trails and named highways. Map keys instructed motorists on which symbols signified which highways so they could see at a glance if they were going the right way.[60] During the war, the DHA began making plans to mark the Dixie Highway route by painting the letters "DH" on red-and-white bands across telegraph and telephone poles.[61] Before the war ended, the familiar DH markers lined the route from Lake Michigan to Miami Beach.

Modern road maps helped to guide military and commercial vehicles during the war and stimulated the tourist industry, but they also served as testaments to the limitations of long-distance automobile routes. Most early auto maps showed only main roads, not the tens of thousands of miles of local roads that fed traffic to routes like the Dixie Highway. Moreover, unlike the old auto guides, modern road maps did not say anything about the condition of the roads they featured. The network of solid lines that appeared on a crisp, colorful Rand McNally Auto Trail map looked nothing like what motorists saw when they drove those routes, especially in the South. To make matters worse, without any oversight of interstate marked-trail routes, it was difficult for mapmakers, automobile businessmen, and highway associations to regulate highway markings. The DHA warned even citizens living along designated routes not to mark the highway without its permission, and they could not stop other groups from claiming portions of the route as their own.[62] By the end of the war, marked trails overlapped one another to such an extent that more than a dozen different markings might appear on a single utility pole.[63] Thus the expansion of the tourism industry, like the war itself, generated new challenges that good-roads boosters proved increasingly incapable of handling on their own.

The rise of the modern tourist industry during the war went hand in hand with the wartime campaign for greater federal funding and oversight for interstate highways.[64] While patriotism justified the Good Roads

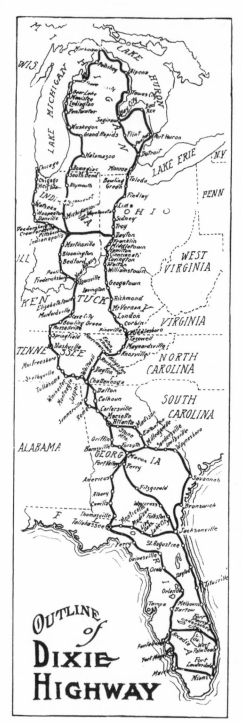

First map published by the Dixie Highway Association, *Dixie Highway*, January 1917, 9.

Red-and-white Dixie Highway markers were usually painted onto telegraph and telephone poles to guide motorists along the route. (Image from Wikipedia Commons, author Parsa)

Movement's wartime push for federal highway aid, profits sustained it. Established businesses and new ones, both national and local, combined to create a formidable advocacy group even more powerful than the original highway lobby. They may have supported the war effort in earnest, but they also understood that the long-term benefits of federal highway funding would outlast the war. Yet in order to convey that message, they could not focus on tourism alone. As they had for years, good-roads advocates needed to make direct political appeals, not only to Congress but also to the local and state officials grappling with the new burden of escalating tourist and military traffic.

Embracing Federal Investment

While good-roads boosters in the United States waited for better roads, American servicemen abroad learned firsthand the value of Europe's much older but higher quality roads. William Weir, a highway engineer who served as an ambulance driver in France during the war, reported that "the original lay out of the French system of highways was fortunately most comprehensive." The old roads, he said, had met "the present exaggerated demands in a most admirable way" and had "undoubtedly been a tremendous factor in the defense against the German invasion." Weir described how French traffic was divided among several different types of roads, including main roads and secondary or side roads; how both maintenance and new construction were performed regularly; and how most roads in the country were already hard-surfaced.[65] In short, French roads embodied everything that the DHA and the highway lobby wanted for American highways, and their service under the most trying of wartime circumstances only further validated the wartime good-roads campaign. The DHA was so impressed with Weir's observations that they summarized his lengthy article, originally written for an engineering journal, in the September 1917 issue of *Dixie Highway*.[66]

Interconnected, hard-surfaced roads like the ones Weir described served not only local traffic and the French military but also American soldiers whose lives depended on good transportation routes. Weir estimated that during the grueling year-long Battle of Verdun in 1916, just prior to U.S. entry into the war, some 5,000 motor trucks had carried troops and supplies night and day along the fifty-mile Bar-le-Duc-to-Verdun Road. Though the French victory came at a high cost, Weir argued that it would not have come at all if not for the influx of new troops and supplies.[67] The DHA predicted that America's fate would also depend on good roads. "Good Roads Saved France," a *Dixie Highway* writer proclaimed just weeks after the U.S. entered the war. "What of America"?[68] Henry Joy, president of Packard Motors and Carl Fisher's old partner in the Lincoln Highway Association, proposed a grim answer by contrasting America's road-building methods with France's long tradition of government-funded roadwork. He praised Napoleon for building "truly efficient roads" with his "absolute power and unlimited resources." French roads, he said, were "the first tentacles of the advance of civilization" and had been since Caesar's day. For more than a century, successive generations had honored Napoleon's commitment to good roads by continuing to devote resources to them, just as Napoleon had honored that of Caesar. In Joy's estimation, if prewar France was at 100 percent efficiency in terms of roads, then the United States was at zero.[69]

Henry Joy was a pioneer in privately planned highway projects, but he had also been a vocal supporter of federal funding for roads and highways for many years. Even before the United States entered the war, he proposed a military border highway that would connect with inland routes like the Lincoln and the Dixie.[70] A few months into the war, Congress considered a similar plan under the Chamberlain-Dent highway bill. The bill, which had the backing of the American Automobile Association and other highway-related businesses, proposed granting the War Department the authority to plan a system of improved highways throughout the nation to facilitate the movement of troops and supplies in times of war.[71]

The Dixie Highway Association supported the bill, not least because it favored the South, with its plethora of military camps. Quoting Will Irvin, a war correspondent on the western front, a Dixie Highway supporter argued that the Chamberlain-Dent Bill was not far-fetched: "'The mechanical transport of automobile trucks on the Western Front . . . has served to counteract for the French, the German strategic railroads. Along with the valor of the men at the front it saved Verdun.'"[72] Good-roads clubs also expressed their support for a federally financed military border highway. In October 1917 the Southern Appalachian Good Roads Association, composed of smaller state and local clubs from throughout the South, voted to endorse the Chamberlain-Dent Bill at their annual conference in Nashville.[73] But backers of the bill failed to generate sufficient support in Congress.

The argument for roads as a military necessity convinced the government to take on more responsibility for funding roads and highways, just not necessarily at home. Even while the Chamberlain-Dent Bill floundered, and finally died, in Congress, the army recruited engineers, contractors, road-machine operators, and surveyors and draftsmen to help build roads and bridges on the western front, where the action was. "Road Builders! The National Army Needs You—*Now*!" one advertisement blazoned. The army hoped to build a highway regiment of more than 10,000 skilled volunteers, not mere "pick and shovel men," to plan and oversee the repair of old roads and the construction of new ones in the most heavily trafficked areas of the war zone. The war showcased skills seldom celebrated at home. Military officials boasted that American techniques were new to European engineers and that the war therefore offered "a great opportunity . . . to demonstrate what skilled Americans can do with modern equipment."[74]

Around the same time, the wartime highway campaign began to have some traction at home as well. In November 1917 Congress acknowledged the severity of the transportation crisis by establishing the Highway Transport Committee (HTC) within the Council of National Defense, an advisory panel President Wilson had set up in 1916 to coordinate the various military and executive departments involved in the mobilization effort. The HTC's job was to figure out how to incorporate waterways, streetcars, and motor trucks into the military transportation network in order to make "the utmost possible use of every facility for transportation" of military personnel and supplies. The federal government even ordered thousands of trucks to be put to use moving military supplies from manufacturers in the interior to ports along the Atlantic coast.[75]

In a stroke of good fortune for the highway lobby and the DHA, the government appointed Roy D. Chapin, president of Hudson Motors, vice president of the Lincoln Highway Association, and close associate of Carl Fisher, as chairman of the HTC and Logan Page of the Office of Public Roads as vice president. Already very familiar with the private experiments with truck services already under way, Chapin set out to replicate these tests on a larger and more-public scale. In late 1917 Chapin arranged for some thirty trucks to drive military supplies from factories in the Midwest to ports on the Atlantic coast. Even the heavy December snows did not prevent the three-ton loads of war materials from reaching the East Coast in good time, despite rough roads. Twenty-nine of the thirty trucks in the convoy made it to Baltimore. Forecasting future possibilities for wartime truck transport, one observer declared: "This convoy is the fore-runner of many others."[76]

Under Chapin, the HTC established the Return Loads Bureau to help relieve freight congestion by ensuring that trucks returned from shipping destinations carrying another load of goods. Organized at the local level and often housed in chambers of commerce or local War Bureau offices, the Return Loads Bureau coordinated cooperative shipping arrange-ments by putting truck operators in touch with businesses in cities on a specified route. Drivers visited the bureau's office in each city to see if anyone needed materials shipped and to streamline hauls to eliminate repeat trips. By centralizing organization under the bureau's services, drivers and shippers could expand their options beyond their earlier practices of simply contracting individuals to deliver goods along discrete routes. This new partnership between motor-truck operators and ship-pers was supposed to reduce the cost per ton by moving goods in both directions, to create more-efficient shipping schedules, and, most of all, to free up thousands of railroad cars for essential wartime needs. Accord-ing to the HTC, it was "a patriotic duty and a war time necessity for all shippers using motor trucks, and motor-truck companies, to cooperate fully with the return load bureaus."[77]

Return Loads Bureaus functioned more efficiently in the North, where better roads and shorter distances made the experiment more practical. Federal experiments in the South were more limited but nonetheless in-novative. Several months into the war, the U.S. Postal Service selected portions of the Dixie Highway in Georgia for an Atlanta-based experi-ment using 182 trucks for parcel-post service among southeastern cities such as Chattanooga and Jacksonville. The experiment was designed to relieve freight congestion around major urban areas, but the DHA inter-preted it to mean much more. The DHA called it "an innovation which has all the ear marks of a business deal." By focusing federal attention on the Dixie Highway rather than the thousands of miles of other nameless rural postal routes covered under the 1916 Federal Aid Road Act, the DHA also hoped the experiment would help to "abolish . . . the 'red tape' connected with the use of Federal and State funds for the highways."[78]

While federal experiments were promising, they were all limited to urban areas. Rural counties in the South were in no position to prove anything, either to the federal government or to the DHA, which pressured them to do more work during the war. They struggled to keep up with the demands of wartime traffic. Any additional tax revenues generated by high cotton prices during the war were eaten up in rising maintenance costs. Initially, some counties took on debt to support road improvements. In April 1917, for example, the same month that the United States declared war on Germany, residents along the Dixie Highway in Rome, Georgia, took on $240,000 in debt to pay for five new concrete highway bridges, new culverts to drain rainwater, and road-grading work along the highway. The county road fund contained less than $140,000 in tax receipts and local automobile registration fees, so the bond issue nearly tripled the amount they usually spent on roads.[79] But the county could not take on enough debt to surface the entire length of its portion of the Dixie Highway, a reality that longtime road advocate and superior court judge Moses Wright could not accept. "While neighboring counties are building the Dixie Highway," he warned, "Floyd County has stopped its work, though it is likely that the federal government will soon want the road for military purposes and this county will have nothing to offer the government."[80] The DHA also used *Dixie Highway* to lobby Thomas County in south Georgia to issue bonds for $800,000 in order to resurface the county's dirt roads. For a county that had no bonded debt and had spent only $500,000 on roads over the previous ten years, the DHA's suggestion was unrealistic. Most counties simply could not take on the amount of debt required to upgrade their portions of the Dixie Highway.[81]

Even while urging counties to go into debt in order to build permanent roads, the DHA tried time and again to secure federal funding to complete the Dixie Highway. The propaganda campaign that the DHA waged in the pages of *Dixie Highway* was only part of the effort. Directors sometimes intervened to convince fledgling state highway departments to apply for their allotments of federal aid under the 1916 highway act and apply it to the Dixie Highway. In May 1917 DHA president Michael Allison wrote to Kentucky governor Augustus O. Stanley and state highway commissioner Rodman Wiley and implored both men to use Kentucky's share to fix sections of the highway in that state. Allison rebuked Stanley and Wiley for the state's impassable roads, which were "holding back the greatest good roads project in the south." A gap in Kentucky's portion of the eastern division of the route, he complained, divided North from South and held up "thousands of tourists [who] are waiting to get through" at great financial cost to the state. In case the argument for tourism was less persuasive just days after the United States had declared war, Allison added, the eastern division through Kentucky was part of a "strategic through highway" that needed to be completed "as a patriotic measure."[82] Allison did not stop there. That same month, director Clark Howell reported that the Dixie Highway between Atlanta and Macon, which he argued was "unquestionably" the most important section of the entire route, was to become the state's very first federal-aid road. However, there had been a delay in procuring the funding and starting work on the section, so Howell suggested that President Allison intervene by telegraphing the chairman of Georgia's new state highway department to request definitive information on the status of the project. Allison did so right then and there, and before the meeting adjourned, the chairman telegraphed back, assuring Allison that the state would begin work as soon as the Office of Public Roads approved the state's application for federal aid.[83] Several months later, the DHA intervened yet again to help get one of the worst roads along the Dixie Highway, the route over Cumberland Mountain in Tennessee, designated as a federal-aid project.[84]

Federal-aid designations meant little without the funding to see them through, so the DHA helped several poor counties raise money to match federal aid. In the South, where highway departments were too new and underfunded to provide the matching funds themselves, the counties had no choice. Laurel and Rockcastle Counties in Kentucky, for example, could not raise the revenue to pay for a five-mile stretch of new construction to complete the eastern division through Kentucky. The DHA convinced the state highway department to earmark $50,000 of the state's federal aid allotment for this stretch of road by promising that the DHA would raise the matching funds. When counties had the funds but lacked the influence with state officials to get federal aid designations, the DHA also stepped in to help. In Nassau County, Florida, where sandy soil made road construction more difficult, the DHA intervened with state officials and procured $25,000 of the state's share of federal aid for the county.[85] Sometimes the DHA raised matching funds through wartime membership campaigns, such as the one it advertised in March 1918. For $5.00 apiece, the subscription ads announced, citizens who joined the DHA could "Aid in the Solution of the North-to-South War Transportation Problem." The ad boasted of the DHA's recent contributions in Kentucky and implied that member subscriptions would also be spent directly on roadwork. Through private subscriptions and supervision of local road-

THE NATION'S BUSINESS
—Is to Win the War!

Our resources are unlimited. We are rapidly building up the best army the world has ever seen, BUT — Of what avail are MEN and RESOURCES without

TRANPORTATION?
Railways — Waterways — Highways

With the added war traffic just begun, the Railways are overtaxed almost to the breaking point. They must have relief.

"It is our bounden duty as a Nation," states the Manufacturers' Record. "to find some way to increase transportation facilities or else the whole transportation business will go to smash."

Waterways transportation is being developed, but the limitations are obvious.

The Bulk of the Burden of Relieving Rail Congestion
MUST REST ON THE HIGHWAYS

The South has been effectively cut off from highway communication with the central northern states by the Cumberland Mountain barrier. Three years ago no travelable road crossed these mountains. Through the efforts of the Dixie Highway Association these mountain barriers have been reduced to less than fifteen miles of new construction and seventy miles of surfacing between Cincinnati and Knoxville, and less than thirteen miles of new construction and fifty-one miles of surfacing between Nashville and Chattanooga.

$300,000.00 of State and Federal Aid has just been appropriated by the State Highway Department of Tennessee to complete the Dixie Highway through the state.

Rodman Wiley, State Highway Commissioner of Kentucky, is preparing to let contracts on last gap of mountain barrier in Eastern Kentucky. The Dixie Highway Association raised $50,000.00 by private subscriptions to equal like amount of Federal Aid.

These victories were the result of continuous effort on the part of the Dixie Highway Association.

The Association needs your co-operation to complete this great war work.

Fill out this Enlistment

Blank. Detach and mail

with remittance of $5.00

without delay

Dixie Highway Association
Headquarters: CHATTANOOGA, TENN.

For the purpose of bringing to a full and speedy completion the Great National Project—The Dixie Highway—the undersigned hereby agrees to pay to the Treasurer of the Dixie Highway Association the sum of

--DOLLARS

for_____annual memberships (at $5.00 each) in said Association. Fifty cents from each membership shall be in payment of one annual subscription to "The Dixie Highway."

County.   Signed_____

_____City or Town_____

Date_____191___ State_____

Make Checks payable to W. R. Long, Treasurer, Chattanooga, Tennessee

---

Your membership will Aid in the Solution of the North-to-South War Transportation Problem.

DIXIE HIGHWAY ASSOCIATION
═══════National Headquarters═══════
CHATTANOOGA, TENNESSEE

M. M. Allison, President

EXECUTIVE COMMITTEE

W. R. Long, Sec.-Treas.

M. M. Allison, Chattanooga
Carl G. Fisher, Indianapolis

Clark Howell, Atlanta

W. R. Long, Chattanooga
Richard Hardy, Chattanooga

DHA membership blank, *Dixie Highway*, March 1918, 13.

work, the DHA helped both county officials and the federal government to do their part for the Dixie Highway and the war effort.[86]

The DHA may have begun its ambitious highway project with only the help of county governments, but as the war dragged on, the organization stepped up its solicitations for state and federal aid. A cartoon used in several issues of *Dixie Highway* in 1918 prominently featured an armored tank called "Federal and State Aid" forging a good road where mountainous terrain and bad roads once had impeded travel. Running alongside the tank in cheerful cooperation was a man representing the Dixie Highway Association. Situated around this picture were scenes depicting the important roles that the war, tourism, and individual efforts played in completing the Dixie Highway. And at the top, a hand called "Dixie Highway" shook the hand of the "U.S.," which proclaimed: "My hand, Sir. You have carried on preparedness long before the war, and have saved the day now."[87]

Nationally, the idea of federal support for major highways and not just post roads gained an increasingly wider circle of advocates thanks to the war. The American Automobile Association, perhaps the highway lobby's oldest and most vocal supporter of federal aid for automobile highways, shared the DHA's belief that the federal government could better serve the development of large infrastructure projects than local or private interests alone. Convinced that restrictions limiting federal funding under the Federal Aid Road Act of 1916 to rural post roads kept even federally funded road projects too local in scope, the AAA proposed to Congress and President Wilson that the act be amended to allow funding for military and commerce routes. In his appeal to the federal government, AAA president George C. Diehl carefully downplayed the commercial aspect of national highways while playing up the military value of projects like the Dixie Highway. "[T]he great possibilities of main artery highways," he said, "are becoming apparent to every citizen who wants to see the war brought to its earliest conclusion through employing every source of cooperation in our great country." George P. Coleman, executive chairman of the American Association of State Highway Officials, agreed. "That the effective conduct of the war demands immediate attention [to] proper construction and maintenance of the highways of the country," Coleman said, "is a plain statement of fact."[88]

■ ■ ■

Cartoon, *Dixie Highway*, March–May 1918.

Toward the end of the war, congressmen continued to submit proposals for federal highway funding and a federal highway commission. In 1919 Senator Charles Townsend of Michigan introduced a bill for a 50,000-mile federally constructed national highway system. The Office of Public Roads, the AAA, automobile manufacturers, tire manufacturers, cement companies, the National Grange, more than 700 civic organizations, and at least thirty-eight state highway associations were among the large and diverse groups backing the Townsend Bill, but it went nowhere. Perhaps because the Townsend Bill proposed a comprehensive rethinking of the federal role in road building rather than a mere extension of the 1916 bill, even congressmen who supported it felt they should not act before 1921, when the 1916 bill was set to expire. More-pressing issues like the debates over the Versailles Treaty, controversy over the proposed League of Nations, and the first postwar presidential election in 1920 pushed the Townsend Bill off the congressional agenda for the time being. War may have accelerated state and federal interest in centralized highway construction, but the messy business of ending a war stalled decisive federal action.[89]

Still, the cultural landscape of the United States had shifted during the war in ways that created new opportunities for road advocates and auto businessmen alike. During the war, U.S. automobile ownership more than doubled from 3.3 million to 8.1 million. The failure of the railroads, moreover, had created a new market for trucking as well. Some 227,250 trucks were manufactured in 1918 alone, and production would more than double by the mid-1920s. This increase in both the volume and the weight of vehicles on the nation's roads only added to the pressure for federal intervention that the DHA had so masterfully exploited during the war.[90]

The DHA's wartime campaign did not result in the federal highway commission or the interstate military highways that the good-roads supporters wanted, but the DHA did successfully reframe the debate over federal funding for roads. Recasting tourist highways, market roads, and even postal routes as integral pieces of a national system of military highways forced everyone to rethink the scope and purpose of interstate routes. Dozens of state highway departments, new or even nonexistent when the United States entered the war in 1917, emerged to help organizations like the DHA to centralize control over modern roadwork. By the end of the war, every state had fulfilled the requirements of the 1916 Federal Aid

Road Act by establishing a state highway department.[91] Although they were limited in their powers for a few more years to come, state highway departments were the only agencies that could initiate federal-aid projects and therefore were critical links between the Dixie Highway and the federal government.

Yet there were also reminders that the wartime campaign had been self-serving on the part of automobile businessmen and those in the tourism industry. By 1919 the Dixie Highway Association itself seemed to have all but forgotten its campaign publicizing the benefits of improved highways to the nation's war effort. Instead, *Dixie Highway* covers celebrated the sunny beaches of south Florida and other opportunities for tourists in the South.[92]

■ ■ ■

But the army had not forgotten. On a hot day in early July, a War Department convoy of eighty-one trucks and 300 soldiers and officers lined up near the White House to begin the long journey to San Francisco over the Lincoln Highway. The start in Washington was a concession to President Woodrow Wilson, a southerner who wanted the South to be incorporated into the tour, partly to forestall regional jealousies but also to acknowledge the South's wartime contributions. One of the young officers in the convoy was Lieutenant Colonel Dwight D. Eisenhower, who had spent the past two years training troops stateside and, as he later recalled, was "mad, disappointed, and resented the fact that the war had passed me by." Looking for "a genuine adventure," Eisenhower volunteered for the assignment. To the War Department, however, the transcontinental trip was more pragmatic. In addition to being a test run for the new trucks and tractors the military had acquired during the First World War, the army also hoped to fatten its thinning ranks by using the tour to advertise the new Motor Transport Corps, a division of the Quartermaster Corps with the responsibility of acquiring, operating, and maintaining military vehicles. The tour was also the federal government's way of showing its commitment to the Good Roads Movement and, more specifically, to the benefits of trucks and highways to move men and equipment. The army's tour generated as much excitement as popular automobile tours had in the earliest days of the automobile, with crowds and elaborate receptions greeting the convoy at almost every stop. But it also suffered its share of

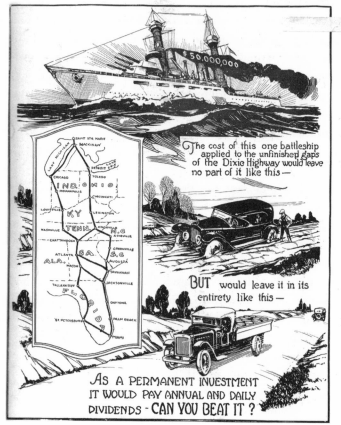

Cover page, *Dixie Highway*, December 1921.

mechanical breakdowns and impassable roads. When the dusty convoy finally rolled into San Francisco, however, all agreed that the roads, not the trucks, needed the most work.[93]

That was the lesson the Dixie Highway Association had been teaching for two years and apparently had not forgotten, after all. The war was not far from the minds of *Dixie Highway* writers two years later, either, when the magazine ran a cover featuring images of a $50 million battleship, a muddy road, and a modern automobile highway. "The cost of this one battleship applied to the unfinished gaps of the Dixie Highway," the caption proclaimed, "would leave no part of it like this"—showing a car imbedded in the mud—"but would leave it in its entirety like this"—showing a farm truck rushing produce to a market. "As a permanent investment it would pay annual and daily dividends," the caption continued. "Can you beat it?"[94]

The Dixie Highway and the Good Roads Movement had benefited enormously from the wartime mobilization, even though the wartime good-roads campaign did not inspire a comprehensive system of federal highways. The end of World War I abroad in late 1918 signaled a new beginning for Dixie Highway boosters and the Good Roads Movement at home. Still, the DHA's use of a battleship cartoon in 1921 was telling. More than three years after the end of World War I, they were still using wartime imagery to argue for federal highway aid. The war provided a watershed of opportunity for road builders, and the DHA had made the most of patriotic rhetoric to mobilize federal dollars. After the war, however, the DHA struggled to identify a peacetime message. Without a national emergency to galvanize support for national highways, the DHA faced the challenge of shifting their message to sustain local and state sources of support for roads.

Modern Highways and Chain Gang Labor, 1919–1924

After World War I, Lillian Smith found herself worrying about money for the first time in her life. Her father had lost his profitable turpentine mills during the wartime economic upheaval and moved his family from their large, comfortable house in north Florida to the family's small vacation cottage in the north Georgia mountain town of Clayton, where they ran a camp for girls. There, Smith and her nine brothers and sisters learned what it was like to scrape by on homegrown eggs, milk, butter, meat, apples, and the few provisions their father bartered for at the local store. Like so many other southerners, they did not benefit from the postwar economic boom. "We were not alone in being poor," Smith recalled forty years later. "Our region was deep in a depression long before the rest of the country felt it."[1]

The children anticipated a gloomy first Christmas in Clayton and wanted to skip it altogether, but their father had other plans. "In that year of austerity," Lillian remembered, "he invited the chain gang to have dinner with us." Smith's father had begun visiting the prisoners in town, where the county housed them in two "shabby" railroad cars when they were not working on the local roads. When he told the chain gang foreman that he would like to host the men for Christmas dinner, the astounded man could only stammer back, "All of them?" Just before noon on Christmas Day, Lillian and her family watched four dozen inmates, some in chains and all in stripes, marching under armed guard toward their tiny house. The children had helped their mother prepare two caramel cakes, twelve sweet potato pies, backbone-and-rice, Brunswick stew, hot rolls, and a washtub of apples—a spread that made her father proud but left him looking pale, "probably wondering what we would eat in

January." Still, he greeted the men warmly, making jokes and inquiring about their families while the children alternately helped their mother in the kitchen and eavesdropped on the curious guests assembled outside on the lawn. When it was time to eat, Smith's father unnerved the prison guards by asking a convicted rapist and a bank robber to go inside and help his wife carry the heavy pots and pans full of food to a serving table set up on the front porch. "Afterwards," Smith recalled fondly, "the rapist and two bank robbers and the arsonist said they'd be real pleased to wash up the dishes." [2]

Later that evening, after their grateful guests had returned to captivity, the children gathered around the fire and listened to their father explain his decision to share what little they had with the county's convicts. "We've been through some pretty hard times lately," he began. "I want my children to accept it all—the good and the bad, for that is what life is. Those men, today—they've made mistakes. Sure. But I have too, Bigger ones, maybe. And you will. Don't forget that." Pausing, he added: "But the one [mistake] I hope you won't make is to cling to my generation's sins. Don't be afraid to change things if you can think of something better."[3]

Lillian Smith's father did not elaborate on his generation's sins, but certainly one close to mind that Christmas Day was the failure to "think of something better" than the harsh penal system that punished misdemeanors and felonies alike with backbreaking labor on the South's dusty roads. Making prisoners work off their debts to society was an age-old practice, but it became a key feature of the southern justice system after the Civil War damaged or destroyed many of the region's prisons and overturned the social and racial order that slavery had long preserved. Along with new segregationist Jim Crow laws and extralegal racialized violence, convict leasing and chain gangs were used to simultaneously restore white supremacy and rebuild southern industry and infrastructure.

As historian Alex Lichtenstein has argued, the transition from the inhumane and exploitative convict lease system that enriched industrialists to county-run chain gangs working the public roads epitomized the South's unique brand of Progressivism. While chain gangs certainly "embodied the brutality of southern race relations, the repressive aspect of southern labor relations, and the moral and economic backwardness of the region in general," they, according to Lichtenstein, were conceived as a "quintessential southern Progressive reform and as an example of penal humanitarian-ism, state-sponsored economic modernization and efficiency, and racial moderation." An old folk saying held that "bad men make good roads," but reformers believed that county chain gangs could rehabilitate bad men and bad roads alike.[4]

But the transition to chain gangs also allowed southerners to embrace modern reform movements such as the Good Roads Movement without sacrificing the racial control—and the local control—that was still central to the region's social, political, and economic institutions. Reformers may have genuinely believed in the capacity of chain gangs to rehabilitate "bad men," but ordinary southern whites cared little for humanitarianism or reform impulses. Chain gangs replaced the hated statute labor system, which exploited white taxpayers' labor with little benefit, with a system that allowed county governments to control and exploit black labor. With the overwhelming support of southern voters, chain gangs did the majority of work on the South's roads between their inception in the early 1900s and their decline in the 1930s.[5]

■ ■ ■

Chain gangs became the most visible symbols of the Good Roads Movement in the South, but they also impeded the work of the movement. Although critical to the construction and maintenance of the local roads that comprised the Dixie Highway, chain gangs recommitted southern states to unskilled labor and uncoordinated, decentralized control over roadwork at the very same time that the highway lobby was working to modernize and centralize it in the hands of state and federal agencies. The racial and political motives that perpetuated chain gangs drove a wedge between proponents of the prison-labor system and Progressives in the Dixie Highway Association, who decried the shoddy roadwork that characterized southern portions of the highway. In place of chain gangs, the DHA promoted expert engineering, paid labor, and modern machinery. To pay for these things, they and their allies backed new sources of funding, such as state gasoline taxes and state highway bonds. Meanwhile, they redoubled their efforts to promote state and federal control, especially after 1921, when a new Federal Aid Highway Act strengthened the state-federal highway partnership. Although their efforts were somewhat successful, southerners' steadfast commitment to the racialized, locally controlled convict-labor system preserved chain gangs as a major resource for southern roads, slowing the work of the highway lobby for years to come and reopening old divisions within the Good Roads Movement at a critical juncture in its history.

Traditional racial politics collided with the modernizing impulses of the Good Roads Movement along the Dixie Highway. As the goals of local whites and highway boosters diverged in the postwar years, they plunged into contentious debates over new technology, state power, and white supremacy. The problems created by a deep commitment to a racialized labor force ultimately undermined the achievements of the highway lobby, the Dixie Highway Association, and white southerners themselves.

Racial Control and Local Control

The practice of rehabilitating prisoners by making them work did not originate with chain gangs. By seeing to it that criminals were "constantly employed," early nineteenth-century state prison reformers in Mississippi, for instance, hoped both to make better citizens out of petty criminals and to distinguish themselves from the vigilante purveyors of the lash and the hangman's noose.[6] State prisons turned into self-sufficient farms and factories that produced everything from food crops to textiles. Because slave masters served as both judge and jury for black men and women accused of crimes, most prisoners in this system were white. After the Civil War, however, new laws targeting everything from drunkenness to vagrancy ensured that black inmates outnumbered whites in county jails and state prison farms in the South. The influx of black prisoners strained the capacity of southern prisons, many of which did not survive the war intact, creating a new problem for southern whites determined to control black men and their labor in the New South.[7]

Southern states found an answer to their problems in the practice of convict leasing, which had been used only sparingly in the antebellum years.[8] Arrangements under the convict lease system were eerily familiar to newly emancipated black southerners, who were singled out for the worst treatment. Under Mississippi's first convict lease in 1868, for instance, planter Edmund Richardson left white convicts behind in the state penitentiary in Jackson and transported only black convicts to his farm in the Yazoo delta, where they toiled without pay.[9] That same year, Georgia issued the state's first convict lease to a contractor on the Georgia and Alabama Railroad who paid $2,500 in exchange for "one hundred able bodied and healthy Negro convicts" to work on the railroad for one year. A few months later, the same contractor leased another hundred convicts to work on the Selma, Rome and Dalton Railroad in north Georgia for just $1,000. Between May and December of 1868, sixteen convicts died on this line, prompting a state prison official to protest the lack of "humane" conditions. Contractors, he reported, often worked the convicts to the point of sickness or exhaustion, then returned them to the penitentiary and demanded a fresh new supply. His complaints fell on deaf ears. Georgia continued to lease convicts, sometimes for free. By the 1870s, the state was issuing leases for up to twenty years, a clear indication of its intention to commit to convict leasing for the long term.[10] As historian David Oshinsky has argued, what began as a "stop-gap measure" to address overburdened prisons lasted long enough to devastate an entire generation of black prisoners.[11]

The value of convict labor and its place in the postwar southern social and economic systems are critical to understanding why southern Progressives later co-opted racialized prison labor during the Good Roads Movement. Controlling black labor was an essential feature of the convict lease system, but southern states also viewed it as an economical way to not only rebuild but also significantly expand the region's industrial infrastructure after the war. Railroads were the first to benefit from systematized convict leasing, as they were necessary for any further economic redevelopment. Georgia was home to some of the region's major railroad hubs, but the state could not offer direct aid to railroad companies as they had in the past. Convict leases proved vital to the early stages of railroad construction because they allowed the companies to build the minimum five to ten miles of track required for receiving state bond endorsements, which in turn helped the companies attract enough private capital to complete the lines. The arrangement mercilessly exploited prisoners, but it enabled southern states to rapidly rebuild their railroads. In Georgia alone, railroad mileage increased 59 percent during the decade following the Civil War and doubled between 1875 and 1890, increasing the state's industrial revenues at little or no cost to the state. Coal mines, sawmills, and ironworks also benefited from convict labor, allowing the region to expand its industrial labor force without drawing on the free labor pool needed for cash-crop production.[12] As David Oshinsky has observed, convict leasing became an "apprenticeship system" for miners in Alabama, half of whom "learned their trade in chains." Their labor allowed Alabama to become the nation's sixth-largest coal producer by 1910.[13]

The convict lease system played such a pronounced role in rebuilding New South industry that early good-roads advocates favored using prisoners to build and maintain roads as well. Experts such as University of Georgia civil engineering professor O. H. Sheffield believed chain gangs were superior alternatives to the statute-labor system, which he derided as "a relic of feudalism and the dark ages." Sheffield argued in 1894 that putting convicts to work on the public roads was "a rational solution of the road question" and solved the state's mutual and "vital" problems of bad roads and bad men.[14] North Carolina's state geologist Joseph A. Holmes called convict labor an "essential feature" of any system of public road improvement.[15] Even Roy Stone, the first chairman of the Office of Road Inquiry, bragged to the secretary of agriculture about the successful utilization of convict road gangs in several states in the South and the West.[16] In Georgia, county governments determined the punishment for misdemeanor convicts, so as convict roadwork became more popular, so did the number of arrests for such trivial offenses as loitering, vagrancy, or, in at least one case, simply looking "suspect." As Lichtenstein has shown, the counties with the first convict road gangs in Georgia were those with the largest black populations. By 1901 the twenty-five Black Belt counties of middle Georgia had 90 percent of the state's surfaced roads and three quarters of the state's graded roads. Overall, nearly 20 percent of roads in those counties were improved, while in the 112 other counties that still depended on statute labor, only 1.4 percent of roads were improved. A county commissioner in one of those counties declared: "I am satisfied that we can never have good roads until they are worked by direct taxation, and the use of the Chain gang."[17]

Convict leasing and county chain gangs shared the prison labor pool for years, but the growing popularity of the Good Roads Movement along with broader Progressive reform impulses put the two on a collision course. Although the two systems exploited prison labor in very similar ways, chain gangs merely relieved counties of the burden of paying wages for road labor, while convict leasing had the added advantage of generating additional revenue for the state. States had few incentives to give up lucrative leases with private companies in order to supplement county road gangs, so county commissioners essentially competed with contractors for convicts. When labor shortages raised the cost of wage labor, more and more contractors turned to the justice system for help.[18] Even counties could not resist the temptation to profit from the labor of their prisoners. They sometimes negotiated profitable leases with local businesses rather than put their convicts to work on local roads. Such contracts were illegal in Georgia, but the practice was rarely punished. In an effort to resolve the problem without abandoning convict leasing, the Georgia state legislature in 1903 offered counties the option of supplementing their misdemeanor road gangs by offering them access to state felons with sentences of less than five years.[19]

By then, the convict lease system was facing a full-on assault. Those who objected to it on humanitarian grounds called leasing "a relic of barbarism" and argued that the punishment of convicts "ought not to be at the hands of a private party who may be tempted by the exigencies of business . . . to make punishment either more or less."[20] Labor reformers called convict leasing "a curse to the laboring millions" and pointed out that competition with prison labor drove down wages.[21] State prison officials—fed up with the illegal county leases that undermined the entire system, not to mention the inability of either the state legislature or the courts to stop them—demanded that the governor do something. By 1908 even the companies leasing convicts had begun to sour on the system. Competition for leases had begun to make leasing less profitable, while several well-publicized mutinies by convicts fed up with their treatment made many lessees skittish.[22]

Under mounting pressure, Georgia's self-styled Progressive governor Hoke Smith called an extra session of the legislature in August 1908 to vote on abolishing the convict lease system. Smith had already kept two of his campaign promises, prohibition of alcohol and African American disfranchisement, but his reform measures were incomplete without a timely solution to the dilemma of the convict lease system. Months earlier, Smith had told legislators that the state's penal system was "unsound in that it too nearly stamps all criminals alike and provides no plan for their reformation." By the summer, he was steadfast in his refusal to even consider any bills to extend the state's current leases, which were set to expire in April 1909. In September state legislators gave in to pressure from Smith and from the public in the face of reports of widespread abuse and abolished the lease system. Following Smith's suggestion that the convicts be put to work "to complete and perfect public highways and other internal improvements," the legislature passed a law putting state convicts under the supervision of counties, where they were to work on local roads.[23] Other southern states also gave up convict leasing, but most continued to use chain gangs for years to come.[24]

Chain gang road labor was employed widely in the Progressive Era United States, but nowhere as extensively as in the South, where, along with disfranchisement amendments and new state primary laws, it comprised a reform agenda based on racial control.[25] Chain gangs exemplified the rationalization of race relations so important to southern Progressives, who wanted to dissociate themselves from the perceived barbarity of both the slave system and convict leasing without sacrificing white control over black labor.[26] They cloaked their defense of white supremacy in distinctly Progressive language. As North Carolina state geologist and good-roads spokesman Joseph Hyde Pratt put it, working convicts on chain gangs "improved their general character and prepared them for better citizenship." Most important of all to Pratt, convict road labor freed the community of "the tramp nuisance," the very people targeted by vagrancy laws.[27] During the governors conference in Chattanooga in 1915, Georgia governor John Slaton used similar language to endorse the use of chain gangs on the Dixie Highway. "When negro convicts are put upon the public roads in the free, pure air," he declared, "I do not know of a more humane way of handling these unfortunate men who must be held in confinement."[28]

There was little confusion among southern reformers that the humiliation of public chain gang labor would be reserved primarily for black convicts, whose "moral standard[s] . . . are not lowered by this form of publicity."[29] Despite their altruistic rhetoric, good-roads reformers were less interested in reforming "bad men" than in making highway work affordable without sacrificing control over black labor. The latter goal was made easier by a number of other powerful forces at work during the 1910s and 1920s, including disfranchisement, the expansion of Jim Crow laws, and exploitative sharecropping and tenancy arrangements that tied rural African Americans to farmwork. The efficacy of these laws, not to mention their role in fueling the mass out-migration of black southerners that began during World War I, is evident in the declining ratio of black to white prisoners in the 1910s and 1920s. Between 1879, when the Georgia state prison began keeping detailed records on convicts, and 1910, black prisoners outnumbered whites by more than nine to one. By 1920 black

Convict gang working on a road in Monticello, Georgia, early 1900s.
(Courtesy of Gary Doster)

prisoners outnumbered whites by nearly five to one, and eight years later, the ratio was almost three to one. Over that same time period, the total number of black prisoners more than doubled, but the number of whites in the state prison system increased by nearly eight times.[30] With the public clamoring for good roads at the lowest possible cost, county authorities assembled chain gangs with at least as much concern for availability as for racial makeup. County road authorities complained that they could not do their jobs without a sufficient workforce. "Your force is on the decline," one Georgia chain gang superintendent groused to his bosses on the board of county commissioners in 1906, "and no prisoners coming in from any quarters."[31] To solve the problem, criminal courts were often used "as feeders for the chain gang," as a contemporary observer said, in order to maintain chain gangs "sufficient in number to justify the overhead charges for equipment and supervision."[32] Although maintaining convict gangs cost money, many counties did not have the cash to pay for free labor and remained committed to convict labor even as segregation and out-migration limited the number of black men available for exploitation.

Even if chain gangs did not control black labor singlehandedly, conspicuous black majorities on chain gangs bolstered white-supremacist logic about the brutal punishment meted out to prisoners. As prominent southern labor reformer Alexander J. McKelway observed during the hearings over abolition of convict leasing in 1908, witnesses so closely associated convicts with black men that they often referred to them as a "nigger[s]," even if they were white.[33] Indeed, the black majorities of most chain gangs contributed greatly to the widespread lack of concern over the conditions under which chain gangs worked and lived. "Since the public demands that the convict road work be operated at the lowest possible cost," one contemporary observed, the employment of "competent" guards and staff for the convict camps was out of the question. "The average county official in charge of such prisoners thinks far more of exploiting their labor in the interest of good roads," he said, "than of any corrective or reformatory value" that forced labor might provide.[34] In an effort to keep costs down and satisfy white voters—the only taxpayers whose opinions mattered to elected county commissioners—chain gangs labored in the meanest of conditions.

The most famous Georgia chain gang prisoner was Robert Burns, whose 1932 memoir, *I Am a Fugitive from a Georgia Chain Gang!*, and the subsequent film version exposed the brutality of chain gangs in vivid detail. Although Burns was a white New Yorker, his experience paralleled those of thousands of black southerners who wound up on the wrong side of the law. Talked into participating in an Atlanta robbery that earned Burns and his two partners just $5.81, Burns was arrested, convicted, and sentenced to six to ten years of hard labor. After the first of two escapes in 1922 (along the Dixie Highway, no less), Burns was sent to the Troupe County chain gang, which had a reputation as the toughest one in the state. Like most, it was also segregated, a feature that to Burns only exaggerated the injustice of the southern penal system. As he was locked into the white bullpen, he watched a guard re-tally the numbers of prisoners on a small blackboard. "I made the thirty-third white convict," he remembered, in a camp of 102.[35]

Burns's description of Georgia chain gangs was so disturbing that it eventually led to the abolition of chain gangs in the state. But in the 1920s, that was a long way off, and the savagery of life in a chain gang was all too real. Convicts were housed in crude buildings with bars on the windows and no protection against heat or cold save the roof and walls of rough wooden boards. Open toilets in the center of the building served as the lavatory. In poorer counties, there were no residential buildings. Instead, prisoners were crowded into "pie wagons," which often were discarded railroad cars like the ones Lillian Smith recalled. Men slept chained together at night and had to ask permission to move. When they did move, the clanking noise of their chains accompanied them everywhere they went. Burns was fixated on his chains. Prisoners, he said, "worked in them, slept in them, were a part of them. The chains could not be taken off unless they were cut off with a hammer and a cold chisel." The chains and shackles weighed twenty pounds and often rubbed painful blisters onto their ankles, an affliction the prisoners called "shackle poison."[36]

Even mealtimes did not provide much comfort. Breakfast was served just after the day began at 3:30 a.m. and consisted of "very bad coffee," a piece of fried dough, sorghum, and a few slivers of fried pork sides. Eight hours later, prisoners ate a lunch of cornpone and a plate of boiled cowpeas, which were often full of sand and worms. Dinner was nearly identical to the dismal breakfast. The menu never changed, except on holidays, when special meals were served, and on Sundays, when the men did not get dinner at all.[37]

With little rest, inadequate food, and no regular showers to clean off the grime from the day before, prisoners worked long days in the hot Georgia sun. Men regularly "got the leather" if gang bosses determined that they had not worked hard enough. Burns recalled one prisoner called Cowboy who had a difficult time "keeping the lick," or keeping pace with the other workers. One day when Cowboy collapsed in exhaustion, Burns watched in horror as a guard kicked him in the face and the stomach, shouting: "There's no rest for Niggers on this chain gang!" Often specific prisoners were singled out for punishment. Burns recalled that on his very first night on the chain gang, six men were selected for beatings. As one lay face down on a bench, the guards brought out "Black Betty," a leather strap "six feet long, three feet wide, [and] one-quarter inch thick." "And with a terrific crash," Burns remembered, "the heavy strap came down on bare flesh with all the strength of the wielder behind it . . . And so it went—one, two, three, four, five, six, seven, eight, nine, ten. Ten licks and the convict, half fainting or perhaps unconscious, was stood up on his feet—blood running down his legs, and one of the guards carried or led him back into the sleeping quarters." There, the convict's dirty, sweat-stained stripes stuck to his wounds and inflamed them, a reminder that the punishment on a chain gang never ended.[38]

The use of chain gang labor thus accomplished multiple objectives. Clearly, it sustained the structural foundations of racial inequality. It also created the illusion of a democratized labor system by allowing *all* white southerners to benefit from public work gangs rather than granting exclusive privileges to promote the private business interests of a few.

But equally important, chain gang labor allowed for southerners to maintain intensely local control over roads during a time when state and federal power were rapidly expanding. Chain gangs remained under the tight control of elected officials, granting local whites nearly complete authority over the labor and resources required to build roads through individual communities. Like disfranchisement measures such as literacy tests and poll taxes, which were locally administered and thus guaranteed local whites not only control over African American citizenship but also significant influence in state politics, county-run chain gangs guaranteed local officials leverage in dealings with the state. Indeed, by institutionalizing chain gangs in 1908, Georgia officials had enhanced county control by significantly increasing the sizes of local chain gangs. In Rome, Georgia, the influx of state prisoners more than doubled the county chain gang between 1906 and 1915 from an average of forty-three convicts in one camp to 109 in two camps. The county even had to hire an assistant to help supervise the work.[39]

Yet as Georgia transitioned from convict leasing to chain gangs, tensions developed between state and local officials. The state prison commission resented sharing authority over state convicts with county officials, whom they did not trust to care for the prisoners. For years, prison officials filed numerous complaints against county officials for mistreating prisoners on misdemeanor county chain gangs, but state law prohibited them from intervening. In 1908, just weeks after the state legislature abolished convict leasing and replaced it with chain gangs, prison-commission officials implored Hoke Smith to grant them the power to appoint the wardens and guards who would oversee prisoners in the various counties, but Smith refused. Smith argued that the new law placed county-run chain gangs under the direct supervision of the prison commission but did "not increase the power of the Prison Commission over the convicts held by the counties further than to provide that the Commission shall make rules and regulations" governing the care of the prisoners. While prison officials could fire employees who did not adhere to those rules and regulations, they could not tell counties who to hire. Smith believed it "wise" to leave such decisions to the counties, who were "better prepared to select in their counties those who will handle the convicts than the Prison Commission."[40]

Local officials guarded their power over chain gangs and grew increasingly resentful of state intervention into county-level decision making after chain gangs replaced convict leasing. These officials believed that they could do whatever they wished with their chain gangs and rejected the meddling of state bureaucrats. One example of conflict between local and state authorities occurred in Floyd County, Georgia, where the chain gang labored on the Rome portion of the Dixie Highway. In 1917 the Floyd County Board of Roads and Revenues accused the prison warden, J. C. Penn, of insubordination for interfering with the county road engineer's plan for working local roads. The board wanted to fire Penn, but the state prison commission objected on the grounds that the county did not have the authority to fire a state employee. In retaliation, Floyd County officials voted to eliminate the county chain gang entirely, leaving the state without a complaint and Penn without a job to do. As soon as Penn formally resigned his post, the county held a special meeting in order to reinstate the chain gang. They also voted to recommend a local road contractor, W. B. Lloyd, to replace Penn as warden. Perhaps to ensure that the state prison commission accepted the recommendation, the commissioners threatened to again abolish their chain gang "on short notice" should there be "any further friction" between the state and the county over prison wardens.[41]

The power struggle in Floyd County illustrates the profound importance of maintaining local control over chain gang labor. By controlling the day-to-day work of road construction and road maintenance, local whites determined when, where, and how roads and highways were completed. Insisting on control over state employees who worked in the county was a clear rejection of state interference in the still overwhelmingly local business of building and maintaining roads. County chain gangs exemplified the commitment of southern whites to a reform agenda while protecting local control. They provided a cheap and steady supply of laborers who would modernize transportation throughout the South at the same time that they preserved the integrity of old political and racial hierarchies.

But the scuffle between Floyd County officials and the state prison commission over control of the chain gang also exposed many of the system's shortcomings. Even as they fought to preserve local control, some county officials admitted that chain gang labor was inefficient. Between 1908 and 1917, the county had been spending nearly its entire annual budget maintaining local roads. A significant portion of what they spent on roads went to maintain the ever-expanding chain gang. In February 1917, as the drama over Warden Penn played out, county commissioners reported that during the previous year, the county had spent over $91,000, or more than 65 percent of its total annual budget, on local infrastructure. Almost $22,000, or nearly a quarter of the total spent on roads, went to maintaining the county chain gang.[42] A significant portion of that money paid the salaries of state employees over whom the county had little control: wardens, guards, and engineers. The county's own engineer was one of the chain gang's chief critics. "[G]o down on the Bluff road," he complained after seeing the county's expenses in 1914, "and see if you can find over $18,000 in work."[43] During the controversy over firing Penn, the chairman of the board of county commissioners, Scott Davis, argued that chain gangs cost Floyd County as much as 25 percent more than free labor would. Not only did the county have to feed, clothe, and house the prisoners, but the work of the prisoners was inefficient. Davis complained that a few weeks earlier, the cost of crushing and spreading rock along the Dixie Highway had cost the county double what it should have. Having to maintain the chain gangs during the rainy winter months, when they did little work, was an additional burden. "I am certain," Davis concluded, "if the people will try contract work they will never be willing to re-establish the chaingang."[44]

Davis's chief motive may have been to undermine the embattled warden, but county records bear out his claims about the inefficiency of local road gangs. Scarcely a decade into the era of the chain gang, it was becoming clear that chain gangs were not the saving grace of the Good Roads Movement in the South. But Davis was clearly in the minority in Floyd County and in the South at large. Their faith in racial control and local control recommitted white southerners to a brutal system of road labor at the same time that they envisioned a new era of modern interstate highways. As they would soon discover, forcing old patterns onto new horizons created a host of unforeseen problems. Debates over funding, technology, taxes, and state power increasingly pointed to the severe limitations of chain gang labor. Ultimately, the Good Roads Movement was forced to reconcile its true intentions and decide whether it would help build modern highways or simply preserve local control and the institutionalization of Jim Crow.

Divisions within the Good Roads Movement

State and federal officials were complicit in perpetuating local control over road construction and chain gang labor. In its first report to the governor in 1919, Georgia's new State Highway Board defended local treatment of prisoners as "humane" and praised the work of the chain gangs as "reliable and satisfactory."[45] In truth, state officials had few other options than to rely on county-run chain gangs because they had little money of their own. The Federal Highway Act of 1916 left counties responsible for the maintenance of all existing roads, including those that comprised most of the Dixie Highway, so the southern states were indebted to county chain gangs for all work done on those roads. Without local, state, or federal sources of funding to pay for experienced labor or modern machinery, chain gangs remained the cheapest road-building resources available not only to county governments but also to proponents of the Good Roads Movement at large.

Early figures on the efficiency of chain gangs seemed to validate them. By 1910, 107 of the state's 146 counties had chain gangs utilizing a total of 4,579 state convicts. A year later, 135 counties used chain gangs. Within five years, the state's total mileage of surfaced roads had more than doubled, and by 1915 the state had more surfaced roads than any other southern state and ranked fifth nationwide. But these statistics overemphasize the efficiency of chain gangs compared to statute labor, which had accomplished very little in the preceding decades. Within a year of ending the convict lease system, Georgians had spent just $2.5 million, or just about a dollar per person and less than thirty dollars per mile, on the state's 82,182 miles of public roads.[46]

Throughout the 1910s and 1920s, the practical limits of chain gang labor showed in the slow pace and poor quality of road construction. Although "improved" mileage had increased since the implementation of chain gangs, the quality of most southern roads had changed little. After more than five years of employing chain gangs, virtually all of the Dixie Highway within Georgia was still made of dirt—difficult enough in good weather but impassable when rain turned the dusty sand-clay

Convict gang helping to build the Dixie Highway, Fitzgerald, Georgia, ca. 1910s.
(Courtesy of Georgia Department of Archives and History, Vanishing Georgia Collection,
Image Ben-263.)

mixture into thick mud. In September 1921, DHA director Clark Howell
complained that both the eastern and western routes between Chattanooga and Atlanta were "extremely bad in rainy weather." The heavily
traveled southeastern section of the Dixie into Florida was also in poor
condition.[47] Three years later, Howell's codirector, William T. Anderson,
observed that little had changed. Citing figures provided by the state
highway engineer, Anderson reported that short stretches of the highway
in north Georgia had finally been paved, but the vast majority of the route
remained a string of dirt roads. In contrast to considerable improvements
in the northern Dixie Highway states, only half of the Dixie Highway in
Georgia had been improved at all. Of that mileage, just over 10 percent
had been hard-surfaced.[48]

Why, despite such a comprehensive and concerted effort to get road-
building projects off the ground, could the Dixie Highway Association
see so little progress? The answer to this question is twofold. First, state
bureaucracies at the time were too weak to take full advantage of federal
resources. The DHA had redoubled its efforts to coordinate county roadwork after 1916, when the first Federal Aid Road Act made funds available
for the construction of rural post roads in every state. For years, DHA of-

ficials had used their influence to try to persuade state highway officials to designate significant portions of the highway as federal-aid projects.[49] But without sufficient funding and supervision from states, they relied on county tax revenues, automobile licensing fees, and small bond issues to make up the state's contribution to federal-aid projects, an arrangement that greatly complicated the DHA's efforts, not to mention those of the states, to take full advantage of federal aid.[50]

The second problem contributing to the slow pace of road construction had to do with chain gangs. Whereas wage labor would enable counties, and therefore the state, to maintain multiple ongoing, interconnected projects and complete them much faster, chain gangs could not be easily restricted and expanded to accommodate new projects. This slowed the pace of roadwork and impeded the state's ability to take full advantage of federal matching funds available through the 1916 Federal Aid Road Act. Furthermore, tired, ill-treated convict road gangs in the custody of guards and wardens could not compete with healthy workers under the supervision of skilled engineers.[51] And as long as counties used chain gangs to perform the backbreaking manual labor involved in traditional road building, they had little incentive—or money—to invest in expensive but far-more-efficient modern equipment. Road-building technology had advanced far beyond the systems being employed along the Dixie Highway, but the commitment to chain gang labor ensured that the pace of construction would be slow and that the quality of new work would be compromised.

Such issues were deeply frustrating to Georgia's first state highway engineer, Warren R. Neel, who did not believe that the meager resources available to most Georgia counties were sufficient to both maintain existing local roads and perform the carefully coordinated upgrades required by new federal-aid projects. A Georgia native and graduate of the Georgia Institute of Technology in Atlanta, Neel began his professional career in Mexico. Between 1901 and 1916, he honed his engineering skills on a number of state and private projects, including the Pan American Railroad. He returned to his home state in 1916 and became the chief engineer of Georgia's brand new highway department when it was established the following year. One of his first tasks was completing the Atlanta-to-Macon stretch of the Dixie Highway, the state's very first federal-aid project. That job and subsequent ones challenged both his engineering talents and his capacity as a supervisor, but Neel proved adept at managing both.[52]

Georgia state highway system in 1920. The system was composed of some of the earliest federal-aid road projects in the state. (Courtesy of Georgia Department of Archives and History, Historic Map File)

In his first annual report to the governor in 1919, Neel complained that the demands of federal-aid roads stretched the capabilities of county chain gangs because counties did not have enough labor or financial resources to work more than one project at a time. Neel estimated that of the nearly $2.7 million in federal aid available to Georgia between 1917 and 1919, the state had only been able to use $38,710.75.[53] Stretching county resources also recommitted the state to the construction of dirt roads, which were cheap to build but expensive to maintain under increasing automobile traffic. Between 1919 and 1920, counties completed 730 miles of federal-aid roads, nearly 70 percent of which were simple topsoil or sand-clay roads.[54]

The frustrations Neel expressed in his report to the governor represented much larger tensions being played out along the Dixie Highway as the South struggled to reconcile tradition with modernity. In his capacity as state engineer, Neel repeatedly found himself caught in the middle of debates over financing and labor, state power and local control, and bureaucracy and community. Time and time again, the need to finance modern infrastructure projects came into conflict with growing anxieties over state power. Although the Dixie Highway Association had lobbied to mobilize federal resources for road construction during the war years, at the local level, southerners began to express concerns about the growing power of state and federal bureaucracies. Simmering beneath the surface of those concerns were deep anxieties over whether or not such distant bureaucrats had the potential to upset the chain gang labor system.

The looming expiration of the first Federal Highway Act of 1916 revived old debates over the scope and purpose of federal aid. Progressives in the Good Roads Movement were dissatisfied with the limitations of the 1916 law and demanded more-centralized state and federal control over highway construction, but conservatives in Congress blocked their efforts to pass sweeping new legislation. Instead, existing state and federal agencies tried to address the concerns by overhauling their procedures for allocating funding. In response to numerous complaints about the federal-aid application process, the newly renamed Bureau of Public Roads streamlined and expedited the federal-aid approval process and also sped up payments of federal matching funds.[55] A reorganization of the Georgia Highway Department in 1920 further eased the transfer of federal aid to county road authorities. These adaptations, as historian Bruce Seely has pointed out, actually forestalled serious consideration of new highway legislation because a functional federal-aid system undermined demands for sweeping changes. The DHA and the highway lobby continued to push for federal interstate highways, but lawmakers upheld the status quo.[56]

But the shortcomings of roadwork in the South left plenty of room for debate over state and federal intervention. State bureaucrats like Warren Neel were tireless critics of decentralized local roadwork. Neel routinely condemned the disparities between poor rural counties and wealthier urban areas that could afford to hire contract wage laborers to supplement convict labor. Of the seventy-five federal-aid projects under way in Georgia by 1919, twenty-two of them, or nearly one-third, were being done with contract labor alone. Another dozen projects requiring skilled labor for paving and bridges were being done by a combination of contract labor and convict road gangs, bringing the total proportion of federal-aid projects worked at least partly by contract labor to 45 percent.[57] The following year, the state proposed a total of eighty federal-aid projects. Sixty-six of these were let to private contractors, twelve were rejected, two were approved but received no bids from private contractors, and just one (simple "Clearing and Grubbing" work) was completed by a county chain gang.[58] Federal aid was not evenly distributed throughout the state's 160 counties. Half of the state's total allotment of federal aid by 1922 had gone to just thirty-eight of the wealthiest counties, or less than a quarter of the total number of counties. Neel did not expect the situation to improve anytime soon. "[T]he amount which will be expended in the smaller and poorer counties in future," he predicted, "will decrease on account of the fact that these counties have already matched as much Federal Aid as they possibly can."[59] Although it was crafted to aid rural counties at the expense of wealthier urban ones, the 1916 federal-aid bill actually had the opposite effect.

Combined with inequitable distribution of federal funds under the 1916 bill, chain gangs only compounded the limitations of roadwork in poorer rural counties. Together, these problems slowed the DHA's ongoing efforts to complete the Dixie Highway and jeopardized the prospect of building a national network of modern automobile highways. Deeply concerned about the impasse that the Good Roads Movement seemed to have reached, the DHA directors and other highway progressives mobilized support for new forms of funding intended to strengthen and centralize road construction. As each of these ideas met with varying levels approval and ire, they highlighted the core challenges facing road advocates as the expiration of the first federal-aid bill drew near.

Among the first proposals on their agenda were state highway bonds. The DHA vigorously promoted what had become known as the "Illinois Plan" for raising state revenues in hopes of securing enough state funding to supplement work on southern portions of the Dixie Highway. In 1919 the DHA praised the efforts of the Tennessee state legislature to pass a $50 million bond, proclaiming that the state's name should "be enrolled at the top of the list of the progressive states of the Union in adopting . . . the Illinois plan which is sweeping the states of the Union like wild fire."[60] DHA officials declared that a comparable bond issue under consideration in Georgia that year had "unanimous" support from voters.[61] The state highway department endorsed tens of millions in bonds, as well, arguing that the infusion of funds would enable more long-term planning, permit the state to make better use of federal aid, and, most significantly, allow the state to hire more contract workers instead of relying on county chain gangs.[62] But expecting voters to take on that much debt and hand the money over to a new state agency was wildly optimistic. In order to even consider state bonds, many states, including Georgia, would have had to amend their constitutions. The proposals went nowhere. By the spring of 1920, all of the Dixie Highway states had either planned state highways or were preparing to do so, but none in the South had bond revenues or other sufficient sources of income with which to build them. DHA chief Michael Allison was furious over the bond failure in Tennessee and blamed conservative politicians for "double cross[ing]" bond supporters at the last minute by withdrawing their support. In a letter to Carl Fisher, Allison claimed to have "broke[n] into the Governor's private sanctuary and called him all kinds of names I could lay my tongue to."[63] Publicly, Allison and the DHA were more judicious. In a magazine editorial, they pointedly warned of the consequences of decentralized funding: "As long as we have county participation in State highway construction, we can never expect a State highway system."[64]

Facing limited success with the bond issue, highway progressives tried to remedy funding issues by convincing the state legislatures to institute or increase auto licensing fees. The fees taxed automobile owners to pay for state roadwork, but unlike bonds, they did not require citizens to take on any debt. This conservative pay-as-you-go approach appealed to state lawmakers in Georgia, who passed a new motor-vehicle tax law in 1919 to replace an older tax that most counties had refused to pay.[65] But the law did not generate enough revenues to permit the state to collect its full share of federal aid. In 1920 the state had collected a total of $1.7 million in auto licensing fees, but it fell far short of the $4.7 million in federal aid matching funds between 1917 and 1920.[66]

When state matching funds ran out, counties that wanted to pursue federal-aid roadwork submitted requests to the state highway department, which then submitted them to the Bureau of Public Roads (BPR). After BPR officials estimated costs and approved the plans, the counties involved had to give the state highway department half of the money so the state could formally apply to the BPR for federal matching funds. Defending the arrangement in 1921, new state highway chairman C. M. Strahan told the governor: "Unless there be some inhibition of law in Georgia unknown to the board, which would forbid the county from thus using the state highway department (a practice which has been going on for two years) such plans should be entirely acceptable to the Federal Government" because "the county is not known to the Federal Government in [this type of] transaction."[67]

The state's dependence upon counties for help with federal-aid projects was both a boon and a burden to local whites in Georgia. While a few wealthier counties initiated most of the state's federal-aid projects, poorer counties continued to control the pace and direction of most of the state's roadwork by relying almost entirely on local tax revenues and convict labor. Even counties that did drink from the federal trough usually counted chain gangs toward their share of the cost. Labor, after all, was an expensive resource. But with only a faint shot at matching federal funds, poorer rural counties held fast to their control over convict labor. What they lacked in real dollars they made up for with the racial capital of white supremacy.

Conflicts between those who supported the use of chain gang labor and those who advocated more aggressive funding for wage labor and machinery divided good-roads supporters. At the state level, it pitted engineering experts and state bureaucrats like Warren Neel head-to-head against other members of the state highway departments, who were often political cronies with divided loyalties. Most of Georgia's state highway commissioners, for example, also sat on the state prison commission. Publicly, at least, they remained committed to the continuation of chain gangs. Writing to the governor in 1919, highway chairman and state prison official T. E. Patterson declared chain gangs "a success." Directly contradicting Neel's evaluation of the superior advantages of wage labor, Patterson concluded that convict labor had "proven more reliable and satisfactory than that of hired day labor." Patterson and his fellow board members even took credit for the work done by state prisoners on county chain gangs, saying that they looked "upon the convict law as a direct state aid to county road building," not the other way around. Although both sides of the debate would claim improved highways as their main objective, the politics of road construction divided those who had once been allies.[68] The state legislature reorganized the highway department later that year and replaced de facto prison commission appointments with gubernatorial nominees, but the divisions remained. Sometimes the state highway department borrowed county chain gangs to work on state-funded highway projects, including sections of the Dixie Highway.[69]

■ ■ ■

The divisions within the Good Roads Movement over chain gangs grew ever more apparent by 1921, when the 1916 Federal Aid Road Act expired. As congressmen debated the future of federal investment in roads, their debates forecast newly emerging ideological differences about the roles of state and federal power. But while concerns over the limits of federal authority had impeded highway legislation in the nineteenth century, these debates over federal power reflected modern misgivings about the growth of new state and federal highway bureaucracies.

During debates in the U.S. House, Kentucky representative John M. Robsion, chair of the House Committee on Roads, invoked time-honored rhetoric to lambast "propaganda" by the highway lobby in support of proposals such as the Townsend Bill, introduced by Michigan senator Charles E. Townsend, which provided for interstate highways and the creation of a federal highway commission. "They want to take care of the joy riders of America," Robsion complained, but "nowhere seem concerned about the farmers getting their products to market or the millions of consumers in these cities having the benefit of these products." Robsion's committee proposed a bill to extend the current federal-aid program. This, he argued, was more in line with "the viewpoint of the road builders . . . and the farm organizations . . . [who] see the great necessity of connecting these rural communities and the producers with the cities of the country," a perspective that protected the states against too much federal oversight. Georgia congressman William W. Larsen, who served on the committee with Robsion, argued that proposals such as the Townsend Bill would "deprive the States of self-government" by giving federal highway commissioners the right to select the roads on which federal aid could be used. "These men," he said, "with little or no knowledge of local conditions, and perhaps caring less, are to be empowered to say what roads shall or shall not be constructed in Georgia and other States. Why should the Government be given the supreme authority?" Larsen also looked out for the interests of his rural constituents by making sure the House committee's bill would allow counties to continue to share with states the responsibilities of selecting and funding federal-aid projects. With overwhelming support from southern congressmen, both the Senate and the House passed bills to extend the existing federal-aid legislation, and it was signed into law in November 1921.[70]

The new Federal Aid Highway Act of 1921 was a bill born of politics and political compromise. The bill was deeply informed by the debates of the preceding years—debates that had revealed emerging political tensions between centralized power and local control. As in 1916, Progressives failed to gain enough congressional support for a centralized federal highway program and had to settle for a more modest bill that exemplified Congress's evolving yet still cautious response to the Good Roads Movement. Like the 1916 bill, it channeled federal matching funds through state highway departments, and the $75 million price tag on the 1921 Federal Aid Highway Act remained unchanged from 1916. But the new law did contain important changes—most notably, a small but significant alteration of the word "road" to "highway" as the destination for federal matching funds. Whereas the 1916 bill designated fund to rural post "roads," the 1921 bill provided funds for limited interconnected state "highway" systems comprising up to 7 percent of the total mileage of roads in each state. The seemingly small change indicated a significant shift in thinking about federal responsibility toward investment in infrastructure. It authorized federal dollars for the construction of state highway systems and effectively granted new administrative powers to state highway agencies as well as the federal Bureau of Public Roads. Yet it did so within strict limits set by the original federal-aid bill in 1916, which mitigated federal authority.[71] By 1921 the politics of highway construction proved to be about far more than just roads; they encapsulated the political growing pains of a nation conflicted over the role of government in a modernizing world.

State Highway Revenue and Convict Labor

The 1921 bill was a mixed victory for state highway advocates like War-ren Neel, who had lobbied for greater state influence in the administration of highway construction. By overturning the 1916 law's emphasis on rural post roads, Congress acknowledged the demand for long-distance highways. And by validating the partnership between state and federal highway agencies, the new bill tacitly endorsed the bureaucratization of highway construction, a move very much at odds with the commitment to local control in the South. But the new bill also underscored persistent problems because it did not solve the fundamental problem of paying for long-distance highways.

Moreover, by authorizing the creation of state highway systems, Con-gress inadvertently magnified the limitations of chain gang labor. After the passage of the 1921 bill, the state highway department of Georgia, for example, approved plans for 5,500 miles of state highways. The roads that comprised the new state highway system included portions of the Dixie Highway, which the state inherited from dozens of county govern-ments. In an early survey, Neel and his team of engineers discovered that 1,000 miles of roads in the new system had been improperly constructed, making nearly 20 percent of the new state highway system more costly to repair. And the funds in hand to maintain the new state highways were already "entirely inadequate," Neel complained, at just $200 per mile. Moreover, all of the roads chosen to become state highways were within a few miles of 80 percent of the state's population, so traffic was heavy and regular maintenance essential. Over 5,000 miles of the 5,500-mile state highway system were still dirt roads, the hardest and most expensive to maintain. Even with matching funds from the federal government, Geor-gia could only afford to start building 1,185 miles, or just one-quarter, of the new state highway system.[72]

Financial limitations limited the scope of state highway work, leaving the state nearly as dependent upon the labor of chain gangs as it had been before 1921. In 1922 the state collected just $1,764,794.67 in automobile licensing fees, its only source of income. Counties, supplemented by federal aid, spent nearly $1 million more that year, and they had three times as much set aside for 1923.[73] The state, however, had far fewer miles to build and maintain. The 5,500-mile highway system constituted less than 7 percent of the state's roads per the guidelines of the 1921 law. County road authorities, with two or three times the funding but some fourteen times the mileage of the state, had no choice but to continue working most of the state's roads with cheap convict labor.

The compromise at the heart of the 1921 federal-aid act failed to fix many of the fundamental problems inherent in road construction throughout the South. It created a more powerful and streamlined state bureaucracy but left local control over chain gang labor intact. The extension of federal aid to state highway systems was hardly a challenge to the authority of county governments, who maintained firm control over state prisoners and the vast majority of roads within the state. Georgia's proposed state highway system was nearly as large as the entire Dixie Highway network, yet it comprised a tiny fraction of the public roads in the state. With little direct state aid for highways, chain gang labor remained a vital part of Georgia's road program. Following the institution of the state highway system, the sizes of county chain gangs remained stable.[74]

. . .

After 1921, state and local figures continued to experiment with new solutions to timeworn problems. Georgia's highway department had for years been trapped between county governments, who struggled with the awesome responsibilities of roadwork, and state laws, which limited the department's ability to take control of its own highway system. Neel pleaded with the governor and the state legislature to pursue new sources of state revenue for state highway construction as soon as possible "in order that the various counties . . . might be relieved of any further expenditure of funds on the State system."[75] The absence of a sufficient state fund with which to match federal aid, he repeated two years later, was a "handicap" that stretched the limits of county resources and compromised the quality of roadwork.[76] By April 1922, Georgia had only 1,100 miles of federal-aid highways either completed or actively under construction. These were the best roads in the state, yet three-quarters of them were still dirt roads. Even the cheapest hard-surfaced roads cost three to five times more to build, and counties were already struggling to cover the cost of surfacing roads with the ubiquitous sand-clay mixture that typified nearly all "improved" southern roads. In addition to hiring more wage laborers who could speed up improvement of these routes, Neel argued that state funds could be used to purchase modern road-building machinery such as rock crushers and rollers to help surface heavily traveled state highways. These machines could do the backbreaking labor usually reserved for chain gangs faster, more accurately, and with more uniform results. Over time, road machines were more economical as well.[77]

Like Neel, the Dixie Highway Association characterized centralized state aid as a benefit to counties rather than a challenge to their authority. The DHA expressed hope that the 1921 law would encourage new forms of direct state aid for "interstate and intercounty highways" in the South instead of the decentralized and "haphazard selection of roads" of the past.[78] They supported investment in the advanced technologies that Neel advocated, and they had seen the advantages that modern, hard surfaces had over dirt roads in other states. In 1924, thanks to state highway bonds, the entire highway through Indiana, and much of it through Illinois, was surfaced with concrete, macadam, stone, or gravel. In Ohio, a 200-mile stretch from Cincinnati to Toledo was also hard-surfaced. Farther south, fewer miles were paved, but states were making an effort. In North Carolina, which had been added to the Dixie Highway along with South Carolina in 1918, about 40 percent of the state's seventy miles of the Dixie was either already surfaced with asphalt or concrete or under construction. In Georgia, although most of the Dixie Highway was surfaced with dirt, much of the hard-surfaced or gravel portion of the state highway system was part of the Dixie Highway. The DHA led the way in the quest for modern roads well before highway departments had been established, and they continued to lobby for hard-surfaced roadways through state avenues.[79]

The DHA's allies in the South wanted to follow the example of the northern states on the highway, but state bond issues remained unpopular in the South. Although southern counties regularly supported *local* bond issues to fund *local* projects, they summarily rejected the idea of going into debt to the state. Southern voters—in particular, rural white voters, who held a disproportionate amount of power in the franchise thanks to the county unit system—would fund projects to support their communities but would not permit the state to use money for project beyond their control.[80] Increasingly after 1921, Neel found allies among his colleagues in the state highway department, but bureaucrats were no match for voters and lawmakers. In 1923 Neel and the highway department backed an unprecedented proposal for $70 million in state highway bonds, but the legislature refused to touch it.[81]

Hostility to state bonds again forced advocates for state highway departments to search for alternative sources of funding that would mollify the deeply suspicious political tendencies of southern voters. The best solution they found was to raise revenue through gas taxes. Like auto licensing fees, gas taxes funded a conservative, pay-as-you-go system. Georgia passed the state's first one-cent gas tax in 1921 with little controversy, becoming one of just thirteen states to do so, most of them in the South and West.[82] In 1923 Neel and his colleagues in the state highway department officially proposed to the state an additional two-cent tax on gasoline and motor oil, a 300 percent increase in the state's gas-tax revenues. Half of the proceeds raised would supplement the state's highway fund, and the other half would reimburse counties for the money they had already spent building and improving the routes absorbed into the state highway system.[83]

Tripling the state gasoline tax was an easier sell to Georgia's white rural voting majority than other options had been. Voters had rejected unpopular measures like state bonds because they perceived them as little more than debts incurred by taxpayers. But by proposing gasoline taxes, the state highway board expressed an understanding that Georgia's roads were not used exclusively by Georgians.[84] As Highway Chairman John Holder explained to the Macon Kiwanis Club in March, tourists "do considerable damage to our roads" and should help pay for maintaining them. Holder insisted that the new three-cent tax would not be an unfair burden on rural taxpayers, either, because the heaviest in-state burden would fall on wealthier urban counties, where more people owned cars and whose citizens "would not object, but . . . feel amply repaid by the better roads all over the state." Holder argued that "people love to be taxed when it is a paying investment."[85] For some rural Georgians, however, the appeal of gas taxes evoked lingering sectional tensions. Although rural counties hoped to benefit from increased business traffic through better road construction, they were still loath to spend money roads for the leisure class. Gas taxes, voters believed, ensured that *northern tourists* would help to pay for road maintenance. Representative Culpepper of Fayette County went so far as to try to divert some of the gas-tax funds to past-due Confederate pensions in order to "let the Yankee tourists and joy-riders pay our debts to the gallant old soldiers." Although the legislature defeated Culpepper's scheme, his sentiments about the distribution of the tax burden were widely shared. In August 1923 the Georgia state legislature approved the three-cent gas tax by the wide margin of 120 to 62.[86]

The new tax was the first serious indication that state lawmakers were willing to aid good-roads advocates, county commissioners, and the state highway department in their slow and steady efforts to centralize control over the construction and maintenance of roads. As the gas-tax bill's sponsor, B. F. Mann declared, it was the "most forward step the state has yet taken" to build better roads.[87] Within a year, the tax raised over $1.2 million for highway work in Georgia, a third of the highway department's total budget. By 1925 proceeds from the gasoline tax had doubled to over $2.3 million and constituted 40 percent of the state's road fund.[88] Gasoline taxes quickly became a popular means of financing state highways nationwide. By 1924 thirty-six of the nation's forty-eight states had passed gasoline taxes, including every state in the South.[89]

But the gasoline tax did not generate immediate or substantive improvements in either the state highway system or the tens of thousands of miles of roads still under direct control of county governments. Neel admitted as much in a candid assessment of the department's efforts to monitor maintenance of state highways in 1924. An increase in motor-vehicle tag receipts over the previous year "made it possible to increase the number of patrol sections and purchase considerable new equipment," but patrol sections remained "entirely too long" and the condition of much of the state highway system too poor for the department to keep many roads in "travelable condition" in bad weather.[90] To make matters worse, in 1923 and 1924 Congress temporarily reduced the amount of federal aid available to the states, appropriating just $50 million in 1923 and $65 million in 1924.[91]

This did not bode well for the Dixie Highway. By 1924 the state was maintaining virtually the entire length of the Dixie in Georgia. Neel estimated that the state would end up spending over $6.7 million for the 1,168 miles of the highway in Georgia. Much of the route was part of the state highway system, and the rest was composed of federal-aid projects that the state was required to maintain. However promising the dollar amounts, though, the state's progress on the highway was slow. Most of the mileage that had either already been completed or was under construction consisted of dirt roads. The state had not begun work on half of the highway, and just $600,000 worth of work was actually under way. Yet Neel claimed the state would spend two or three times that amount on the Dixie Highway in the coming year.[92]

DHA director William Anderson openly challenged such an optimistic assessment of the state's ability to improve and maintain the Dixie Highway, much less the majority of unsurfaced state roads, with existing revenues. The gasoline tax and motor-vehicle licensing fees brought in $6.5 million a year, but Anderson estimated that "only one million dollars of that was devoted to highway construction." The state highway department still spent most of its money maintaining dirt roads. Anderson estimated that it cost the state $300 per mile annually to maintain dirt roads, while paved roads would cost closer to $70 a mile. Not only were dirt roads more expensive to maintain, he argued, but all the money spent on them had to be spent again the next year. Every penny of the tag money, he concluded, "is swept into the gutters by every rain annually." Nearly all of the state's gas-tax revenues went to expenses other than road construction or hard-surfacing. Anderson complained that the highway department paid one-third of the gasoline tax proceeds "to oil inspection and to the oil companies and [to] the general treasury" and an equal amount to the general treasury for the baffling "payment of some rental hypothecations of the state road." The rest of it went directly to county governments, but according to Anderson, "lots of these counties spend it on local roads and don't put it on the State Highway system." Because federal aid could be used only for new construction and state aid for either the state highway system or maintenance of federal-aid roads, counties had good reason to divert this modest state supplement to help maintain some of the tens of thousands of miles of dirt roads for which they and their chain gangs were still solely responsible.[93]

Although by 1924 local, state, and federal authorities shared the financial burden of building roads, limited funding from all three, coupled with restrictions on the use of state and federal aid, slowed road improvement to a snail's pace. But for the first time in the history of the nation's roadwork, there was no question that the responsibility for the nation's roadways lay with not only county commissioners but also newer state highway organizations and the federal government. Relations among authorities in each of the three levels of government had improved, as well, with Warren Neel boasting in 1924 that "relations between the Bureau of

Public Roads and the State Highway Department have improved steadily until the point has been reached when I can say that the two departments are working in complete accord." In just five years, Georgia's state highway department had established itself as a key part of the road-building process.[94]

But the limits of that collaborative relationship in the South ensured that the Dixie Highway Association remained a vocal part of the road-building business. In November 1924 William Anderson told the DHA that, although the DHA had no official relationship with the state legislature, he had visited fifty or sixty counties and informed people there of "the actual status of their road money" and "urge[d] them to impress on their representatives their desire to have a different road program so Georgia may have some roads." His idea, he said, was "to issue bonds or double the automobile tax charges" and "to create a sentiment through the Dixie Highway Association, and through every avenue, to get the people to thinking of the enormous loss they are suffering by the lack of roads." Anderson practiced what he preached during a trip to Savannah for a DHA meeting, when he noticed that neighboring Screven County had a thirty-mile stretch of sand road that needed surfacing. When county commissioners objected to the cost, Anderson took his case to the public and even sent an engineer to help the county figure plans for surfacing the road. Anderson was perhaps overly optimistic that Screven officials would follow through, but his real goal was simply to stoke public support for more-durable, and more-expensive, main highways. "We are not satisfied," he said, "and are not going to be until Georgia has her system paved."[95]

The state's regard for the DHA's leadership over the previous decade was evident in Governor Clifford M. Walker's decision to appoint Anderson to the state highway board in 1923.[96] This gave Anderson direct access to both the other members of the highway board and Governor Walker, to whom he appealed for substantial changes in the state's highway laws. In early 1925, Anderson and Stanley Bennet, his colleague on the highway board, wrote to Governor Walker pleading for more money for better roads. "We think there should be a bond issue," the men wrote, "of sufficient amount to enable the department to take up a paving program on our most-used roads." Traffic on these dirt roads, they reported, was too heavy to make maintenance economically feasible. Noting that New York State spent $600 per mile on its dirt roads while Georgia spent about

$250 per mile, Anderson and Bennet warned that maintaining dirt roads would only become more expensive as traffic increased.[97]

But the modernizing designs of progressive road advocates did not have the political support of the people they needed most. Neither Governor Walker nor state legislators were willing to go as far as Anderson, Neel, or the DHA wanted. Hard-surfacing required not only more labor but also more machinery, both of which were costly up front, even if they saved money in the long run. The three-cent gasoline tax had been helpful, but it hardly signaled the intension of lawmakers to green-light the paving of the entire state highway system. It had given the state highway department just enough money to become a legitimate bureaucratic institution but not enough to do the work it needed to do. The reality of the postwar years revealed that state highway men were empowered to play politics but not to build roads.

Throughout the 1920s, the politics of race continued to shape the terms of debate. Southern politicians' commitment to racialized labor haunted road-building policy, even as public sentiment began to sway. In 1922 Neel had reported that he was "receiving letters . . . asking why we do not start [road]work and give relief to so many unemployed people who are actually suffering throughout the state."[98] But once chain gangs had been firmly established in the South, most southern good-roads advocates never considered abolishing them. Instead, they celebrated statistics that ignored the sweeping inferiority of southern roads and highways. By 1923 Georgia led eight southern states in federal-aid mileage, with more than 1,200 miles of federal-aid projects completed. Nationwide, the state ranked first in the value of federal-aid bridges and second in the mileage of federal-aid roads completed.[99] But all of these figures obscured the fact that Georgia's roads, including those within the state highway system, were built piecemeal and county by county as local funds allowed. Northern states, meanwhile, had managed to build continuous hard-surfaced highways. The entire length of the Dixie Highway through Illinois, and much of it in Ohio, was paved by 1924.[100] That year, more than 5,000 convicts worked Georgia's patchwork dirt roads.[101] Georgia held onto most of its deceptive records throughout the 1920s, proving to some at least that the state could be a leader in roadwork even while doing it all "economically" with county funds and convict labor.[102]

Eventually, politicians realized that the best way to resolve the debate between local control and state power was to embrace the very issue creating the rift. If racial control sat at the heart of counties' anxieties about state power, then chain gangs offered the solution to political wrangling. In August 1924 the Georgia legislature approved a bill authorizing the state highway department to begin using its own chain gangs and also to hire guards, wardens, and physicians to supervise and care for the prisoners. For white voters and politicians, it was a perfect compromise. The state had no intention of challenging county roadwork by siphoning off large numbers of state prisoners for its own convict-labor force. In a clear recognition of local authority, the state gave counties the option of loaning the state their own chain gangs, and it empowered the highway board to hire additional guards and purchase extra equipment to support the borrowed convicts at the state's expense.[103] Thus the legislature authorized the state highway board to spend vast amounts of money to expand and sustain the use of convict labor, even while allowing counties to retain control over chain gangs working both local roads and state highways.

Armed with dubious rankings and partial statistics, politicians argued that chain gangs were not only acceptable substitutes for wage labor and modern machinery but also vital components of both local and state road-building programs. Of course, none of the official figures from state or federal highway authorities reflected the blood and sweat of thousands of convicts who had labored on roads for the previous fifteen years. Though they loomed large in the history of modern roadwork, their very existence belied the progress that both public and private good-roads advocates claimed. By securing a stable workforce for federal-aid projects, state highways, and county roads, southern whites prioritized racial supremacy and deployed black labor in the name of southern progress. This may not have been evident in their own statistics, but it was front and center in those of the Dixie Highway Association, who in 1925 reported that of an estimated $1 billion spent on the nation's roads that year, 60 percent still came from county governments. In the rural South, the proportion was probably even higher.[104] The price of progress in the chain gang South was measured not in reliable statistics, as it turned out, but one slow dirt mile at a time.

CHAPTER

· 5 ·

Paved with Politics

Business and Bureaucracy in Georgia, 1924–1927

When Carl Fisher complained in 1912 that highways were "built chiefly of politics," he was referring to the problems that dictated the slow pace of roadwork under decades of county control. Nearly fifteen years later, Fisher's tongue-in-cheek protest could have described the new challenges facing state and federal highway agencies in the modern automobile era. Good-roads advocates had achieved many of their goals in the intervening years, including the passage of federal highway legislation in 1916 and 1921, the implementation of new taxes to raise highway revenue, and continued progress along the Dixie Highway. But as debates over both chain gang labor and the extension of federal highway aid in 1921 had already shown, white southerners viewed the creation and expansion of a new highway bureaucracy with a mixture of enthusiasm and caution. In the 1920s, politics and roadwork were snug companions despite, or perhaps because of, the Good Roads Movement's success.[1]

Nowhere was this more evident than during election season in the South, when roads dominated political debates. As one historian has observed, roads were "a great political football, patronage source, power lever, and an issue that was safe and sure." Politicians exploited the region's bad road situation in order to win favor with voters, "seiz[ing] upon" it "with a religious fervor, as though it would be the salvation of their people."[2] Yet it was not enough for a candidate to convince voters that he was a good-roads disciple. Those aspiring to political office also had to demonstrate their fluency in Progressive Era southern politics. Good politics in the good-roads South required skillful maneuvering in order to guarantee constituents the roads they needed without upsetting the political status quo.

In the gubernatorial campaign of 1926, Georgians confronted the limits of this arrangement. The bitter Democratic primary race pitted state highway chairman John Holder against wealthy physician, planter, and textile-mill owner Lamartine G. Hardman. Both were vocal supporters of the Good Roads Movement, but the similarities ended there. Hardman was an archetypal southern candidate, a wealthy elite with deep ties to the state Democratic establishment. On the issue of funding for good roads, Hardman was a strict proponent of the pay-as-you-go plan, opposed to bonded indebtedness, and cautious about additional taxation for major new construction.[3] As highway chairman, Holder had supported multimillion-dollar state bonds, as well as new taxes, to finance a more-aggressive, centralized state highway agenda. Although he was a former state legislator, Holder was better known as an antiestablishment bureaucrat who skillfully wielded his patronage powers as state highway chairman over legislators desperate for better roads within their districts. As the race for the governor's mansion turned into a referendum on roads, voters had to take sides in a contest that determined not only the state's political future but also the future of road building in the state. By pitting modernizing reforms against traditional social and economic policies, the campaign between Lamartine Hardman and John Holder exemplified the ideological divide that confronted the Good Roads Movement in the 1920s.[4]

While chain gang labor had already compromised road building in the South, the growing power of state and federal highway agencies caused many southerners to reexamine their commitment to the Good Roads Movement. Southern voters began to equate that consolidation of power with the erosion of local control. The highway lobby's successful campaign to convince Congress to fund the first U.S. highway system from 1924 to 1926 further heightened fears of too much centralized state and federal power. In the context of the so-called Roaring Twenties, when business was booming but most rural southerners were hurting, those fears were potent. At the very same time that good-roads advocates received the long-distance highways they had wanted for years, the modernizing project represented by the Dixie Highway derailed.

Farmers and the Origins of the U.S. Highway System

Nearing its tenth anniversary in 1924, the Dixie Highway Association looked back over a decade of remarkable achievements. Congress had passed two comprehensive federal highway bills, state highway departments were building their own highways, and much of the nearly 6,000-mile network of roads that made up the Dixie Highway had been improved to DHA standards and marked with signage. Although most roads in the South, including the Dixie, were still crude dirt roads, tourists and residents could navigate the region more easily than ever before. Between 1915 and 1920, car registrations in the South had increased nearly sixfold to over 146,000. Automakers noticed these developments as well. In the early 1920s, they began heavily marketing cars in the South for the first time. The first dealerships appeared in cities like Greenville, South Carolina, and Chattanooga, Tennessee.[5]

Changes in highway administration went hand in hand with advances in highway work. As World War I was ending in 1918, the Office of Public Roads underwent a transformation. Congress granted the new Bureau of Public Roads (BPR) both more money and more authority, a testament to the significant progress in road building in the quarter century since the agency was first created. Under the direction of former Iowa state highway engineer Thomas MacDonald, who took over in 1919 following the death of Logan Page, the BPR grew into a formidable federal agency. The BPR coordinated the efforts of all forty-eight states to designate state highway systems following the 1921 federal-aid act, and by November 1923 MacDonald had approved every state system and published a map of the entire federal-aid highway system. Although they accounted for less than 6 percent of the nation's roads, even less than the 1921 act had approved, federal-aid roads were within ten miles of 90 percent of the nation's population.[6]

Despite significant progress in integrative road building, the rural South remained on the fringes of the national economy in the Roaring Twenties. The federal-aid system that had been conceived in 1916 to integrate rural Americans ended up serving primarily the people who had opposed it. Farmers still found themselves cut off from new markets and marketing options by poor local roads. Ninety-four percent of city dwellers lived directly on the system, while miles of rough, impassable roads lay between most farmers and the nearest federal-aid road. And because much of the federal-aid system was not interconnected, even farmers who could reach federal roads might not be able to travel very far. Even state highways, local roads linked to form intrastate networks, resembled a patchwork of uneven-quality roads that varied from county to county. Few were connected across state lines.[7]

While urban Americans prospered in the interwar decade, isolated, cotton-dependent farmers in the South waged war with the land. Although farmers had been eager to be on the Dixie Highway in order to expand their marketing opportunities, diversification had not saved the rural South. Even in places where generations of overuse had not depleted the soil, crop yields varied from year to year depending on production costs and kept prices in a constant state of flux. Farmers trying to stabilize their own finances by growing more cotton, whether to take advantage of a spike in prices or to make up for low prices with larger yields, succeeded only in destabilizing the agricultural economy. The vast majority of southern farmers—small landowners, tenants, and sharecroppers—remained trapped in a vicious cycle of debt.[8]

The federal government's response to the agricultural crisis revealed that federal lawmakers did not comprehend the relationship between the rural South's poor roads and its poor economy. Ignoring decades of failed efforts from agricultural reformers to persuade farmers to diversify their crops in order to better weather price fluctuations, the U.S. Department of Agriculture's Cooperative Extension Service continued to tout crop diversification as a remedy for the farmer's ills. The Smith-Lever Act of 1914 enabled the USDA to place expert agents in every state and almost every county to run demonstration farms and educate farmers about new crops. Because contracts with landowners almost always required tenant farmers and sharecroppers to grow cotton, extension agents concentrated their diversification campaign on landowners. But marketing limitations hamstrung most landowners as well. Not only were market facilities for crops other than King Cotton few and far between, but those that did exist often were inaccessible. Even the smallest of towns had cotton warehouses, but most had no facilities for marketing grain, hay, corn, peas, or other crops.[9]

Officials in the Cooperative Extension Service could have looked first to their sister agency within the USDA, the Bureau of Public Roads, for solutions to the diversification dilemma, but instead they urged farmers to participate in cumbersome, railroad-based cooperative marketing programs. In South Carolina, where monthly field reports from county agents provide a rare glimpse into the relationships between local farmers and federal policy, railroad car-lot shipments were the centerpieces of the crop-diversification campaign in the 1920s. Cotton marketing served as their model. Cotton merchants bought individual bales from local farmers and combined them for sale to larger buyers or manufacturers. Without local merchants for other crops, agents tried to substitute for the middleman altogether by coordinating large-scale railroad shipments themselves. County agents would mail circulars and rely on word of mouth to announce when and where the cars would be stopping, what kinds of goods they would be shipping, and how much farmers could expect to be paid for those goods. But car-lot shipments failed to facilitate diversification in South Carolina. Only farmers who could walk or drive the distance to a railroad depot to meet the train on the specified shipping day could participate at all. County agents told stories of poor farmers walking for miles to reach railroad depots to sell one chicken on a car-lot shipment, the only way they could get to town when rain washed out the roads.[10] For more-distant markets, it would have been impossible for farmers to sell even what they could carry on foot, much less transport wagonloads of crops, especially if they did not own cars or if muddy trails swallowed their teams and wagons.

Like the farmers they struggled to serve, county agents often complained about the "mud tax" that not only impeded efficient transportation of crops to market but also interfered with extension work. When Claude Rothell became the new agent for Saluda County, South Carolina, in January 1924, he had a difficult time familiarizing himself with the farms and the people there. The rainy winter weather had reduced the local roads to impassable stretches of mud, so Rothell spent more time digging out his old Tin Lizzie than actually talking to farmers.[11] Farmers faced the same problems when they tried to call on the county agent or attend extension meetings. Often agents would cancel meetings and demonstrations altogether after bad weather because they knew that most farmers would be virtually imprisoned by the muddy roads.[12]

■ ■ ■

The needs of farmers helped persuade the USDA to consider plans for a more-comprehensive system of federal highway aid. While the

Cooperative Extension Service supported railroad-based marketing programs for southern farmers, the USDA backed another of its subagencies, the BPR, in an unprecedented plan to expand highway accessibility. With the Federal Aid Highway Act of 1921 still in place, they could neither increase federal funding nor amend federal authority to build roads directly. But they could work within the bounds of the 1921 law to give state highway agencies the flexibility to build interconnected federal highways.[13]

In late 1924, the American Association of State Highway Officials (AASHO) asked BPR chief Thomas MacDonald and Secretary of Agriculture Howard M. Gore to appoint a joint board of federal and state highway officials to designate a system of federal highways. Gore, a West Virginian, and MacDonald, a rural westerner, understood all too well the limits of existing roads. Unlike federal-aid highways, these routes would be part of an interconnected system planned, financed, and maintained jointly by federal and state governments under existing federal highway legislation. It was not an entirely new plan: the auto industry's highway lobby had proposed versions of the same thing for many years. But this proposal did not provide for a federal highway commission or increases in existing federal funding, and it did not require an act of Congress. All of these reasons likely explain why it was the first proposal to have wide-ranging support from rural and urban factions alike. These routes, the AASHO vowed, would have "a conspicuous place among the highways of the country as roads of interstate and national significance."[14]

Although the Dixie Highway Association did not have a voice in the AASHO proposal, the Dixie Highway and other marked trails and memorial highways had impressed both the USDA and the BPR. Whereas in years past the agency had been suspicious of all marked trails, Secretary Gore acknowledged that the public response to "unofficial designations" clearly conveyed "a desire among a very large number of citizens in every part of our country for such a systematic designation of highways." "Numerous highway associations," he said, "have attempted with some degree of success to select and mark certain highways as interstate routes, and practically all of the State Highway Departments have recognized the need of indicating for the benefit of the traveling public main market roads and other roads of primary importance within their respective States." It was finally time, Gore said, "to correlate the various efforts that have heretofore been made by so many different agencies" in order to create an interconnected interstate system that was less confusing to travelers and better maintained by centralized state and federal agencies.[15]

In February 1925 BPR chief Thomas MacDonald approved the plan and signed off on a joint board composed of twenty-one state highway engineers, including Georgia's Warren Neel, and three BPR engineers. Hoping to avoid a lengthy, acrimonious debate among state highway officials over which routes should be included, the joint board purposely kept the routing selection a secret until October 1925, when it announced a preliminary "skeleton system" of just 50,137 miles. They charged state highway officials with reviewing the maps, suggesting any changes, and reconvening at a later date to make the final official designations.[16] Congress did not appropriate any new funds for the new system, so states had to work within the existing limitations of the federal-aid program. Along with the joint board's plan to allow state officials to approve the designations, this prevented any significant challenges to the new highway system on the grounds that it overstretched federal authority and cleared the way for state highway departments to expand their authority over the nation's highways.

As anyone who remembered the fevered Dixie Highway routing contest a decade earlier could have predicted, the joint board's plan to keep federal highway designations as harmonious and efficient as possible unraveled as soon as the routes were announced. Citizens bombarded the board with requests to be part of the new federal highway system. Some who had been left off the preliminary routes, like rural south Georgians along State Highway 35, pleaded to be added to the federal highway system. Others, aware that the announcement had stirred up controversy among residents left out of the system, wrote in to preempt efforts by their neighbors to get the highways rerouted. Georgia state senator H. J. Carswell expressed gratitude on behalf of Waycross, Georgia, residents for the joint board's decision to route U.S. Highway 1 along the Dixie Highway route into Jacksonville. But he pleaded with the board to "hold it as such, and not consent for it to go over the Coastal Highway," as residents farther east were demanding.[17]

Still others relied on old-fashioned booster rhetoric in order to convince the joint board to add them to the federal highway system. Members of the Carroll Club of Carrollton, along State Highway 8, sent the board a seven-point manifesto outlining their case. Besides being "one of the most important cities in western Georgia . . . [and] a county seat of one of the most progressive counties in the state," Carrollton was a major agricultural center already connected to towns to the east and the west by a state highway. Also known as the Bankhead Highway, named after Alabama senator and sponsor of the 1916 Federal Aid Act, John Hollis Bankhead, the highway made Carrollton a natural choice over a rival route twenty miles north through Tallapoosa, which was not yet completed. For good measure, the club added that their route would probably be paved in the "near future."[18]

As both a state highway official and a member of the joint board, Warren Neel exercised tremendous influence in the designation of U.S. highways in Georgia. While highway chairman John Holder was busy launching his campaign for governor, Neel addressed the routing bids that poured into the highway department's offices in the months following the joint board's announcement. Although he kept his political preferences to himself, Neel had defended Holder in the past and shared his desire for a strong and well-funded state highway program.[19]

Neel's handling of his responsibilities during the simultaneous campaigns to select official U.S. Highways and select a new governor in Georgia confirmed his reputation as a progressive road builder and an ally of the farmer. In late 1925 Neel asked the joint board to consider designating an important east-west route through the heart of Georgia's agricultural Black Belt. The proposed route—which entered the state via the Alabama state line and passed east through Donalsonville, Thomasville, and Valdosta and on to the port city of Brunswick—linked valuable, if struggling, farming districts. Making a case for the highway, Neel told the joint board that this was "the wealthiest part of the state" and that, if approved, the route would be "the only [U.S. highway] on the map south of Macon."[20] A few weeks later, commissioners in Colquitt County asked the state highway commission for help in getting a second south Georgia route, a string of county roads known as the Florida Short Route, designated as a U.S. highway. Having another "higher-class" highway maintained by the federal government, the commissioners argued, would be of immense value not only to Colquitt County but to all of south Georgia. State highway officials agreed to help with this route as well.[21] Neel also intervened in the increasingly testy Carrollton-Tallapoosa contest, recommending that the joint board consider adopting both highways as alternate U.S. highway routes and thereby giving more residents in the same part of the state access to major interstate routes.[22]

Although Neel tried to help farmers turn their county roads into U.S. highways, he also helped his own agency by selecting state projects in need of additional funding. Neel had backed successful efforts to impose auto licensing fees and gasoline taxes to fund state roadwork a few years earlier, but he had long complained that these sources of revenue fell short. A longtime critic of the pay-as-you-go plan, Neel reiterated his opposition going into the gubernatorial election year during a speech to the Atlanta City Club in March 1926.[23] Despite his best efforts, however, he had not persuaded state leaders to reconsider their opposition to state bond issues. To help complete state highways, Neel recommended a number of them for federal highway designation. Some, like Highway 8 through Carrollton and Highway 35 through Waycross, were on the joint board's initial maps, which Neel helped devise. But in February 1926 the state highway department requested that two more state routes between Chattanooga and Cartersville, just north of Atlanta, be added to the U.S. highway system. One was State Highway 1 through Rome, and the other was Highway 3 through Dalton, the eastern and western divisions of the Dixie Highway in north Georgia.[24]

Like Georgia farmers, Neel understood the importance of prioritizing interconnected, continuous routes such as state highways and the Dixie Highway. Along with well-maintained local roads, these long-distance highways formed a sophisticated national highway system that served all citizens, not just those in urban areas. He tried to help farmers by ensuring that the new U.S. highway system linked rural south Georgians to not only the rest of the federal system but also to state highways and existing interstate routes like the Dixie. But even if Neel's work pleased some farmers, it was bound to disappoint those who were excluded from the new highway system. To farmers, even interstate highways were local issues. As the routing contest played out alongside the increasingly rancorous gubernatorial race, local politics and highway bureaucrats like Neel clashed over control of the nation's first federal interstate highways.

From Names to Numbers

Even as it grew nationwide and made significant inroads in the South, the auto industry fretted about the slow and difficult work of transforming rough local roads and disconnected state highways into interstate highway systems. In August 1925 the president and general manager of the American Automobile Association commenced a cross-country drive from Washington, D.C., to San Francisco. Driving a powerful Cadillac touring car, they hoped to restore public faith in motor travel by proving that modern cars could handle long distances and rough roads with ease. Halfway into the trip, however, the two men got lost. Numerous marked trails led west, one of them recalled in his journal, "and each road was worse than the other." Not sure which route to take, the motorists selected one and spent the next two hours bouncing over an almost impassable road, "dropping into chuckholes and raising clouds of dust." Their quest to prove the reliability of new cars had, like similar adventures had so many times before, only proven the inadequacy of old roads. However, this cross-country trip also exposed the shortcomings of years of uncoordinated road building. These AAA officials would be among the last tourists to travel the country's marked trails.[25]

By 1926 the systemic limitations of privately organized, locally built roads began to give way to the expansive possibilities of federal highways. In October 1925, the same month the joint board unveiled the preliminary U.S. highway map, the Dixie Highway Association's board of directors disclosed that the group was facing "a more or less serious financial situation."[26] The following summer, the DHA's executive committee met to discuss not only the organization's financial affairs but also its future. They decided against disbanding the organization, but with advertising revenues down, DHA officers voted unanimously to suspend publication of *Dixie Highway*.[27] With its most effective mouthpiece silenced, the DHA struggled to be heard in the crusade to replace unofficial named highways with numbered federal highways.

The DHA's troubles reflected the changes that had taken place in highway work over the past decade. The rate at which Americans were purchasing automobiles nearly doubled between 1921 and 1925, adding to the pressure on state and federal highway officials to prioritize the most heavily traveled routes.[28] Many of these were named highways. Like the Lincoln Highway and others, significant portions of the Dixie Highway had already been designated as federal-aid roads and/or state highways. For instance, by 1921 the Dixie Highway through Rome was also known as Federal Aid Project 14. Just south of Rome in Paulding County, the Dixie overlapped with both Federal Aid Project 28 and State Highway Project 9, later State Highway 6. The southeastern division of the Dixie Highway through Waycross was also State Highway Project 9 and two separate federal-aid projects. In all, no fewer than forty-four sections of the Dixie Highway in Georgia became federal-aid highways. Nearly half of the total state highway and federal-aid mileage in the state were part of, or connected to, the Dixie Highway.[29] By the mid-1920s, marked trails, federal-aid roads, state highways, and local market roads overlapped in a modest and often disjointed but vital network of high-traffic roadways.

The joint board had chosen many of these same routes to become U.S. highways, and these federal designations would further marginalize marked-trail associations. For years, the DHA had looked to state and federal agencies as their allies in the Good Roads Movement, but public highway officials still viewed some marked-trail associations with suspicion. Soon after approving the AASHO's plans to designate a federal highway system, USDA chief Gore remarked that he had done so in direct response to the proliferation of unofficial highway associations. While some groups were legitimate, many of them, he said, "have no particular significance of any character" and go around "making collections under the guise of membership fees from people along the way." It had grown so bad that the AASHO was compelled to take over the main highways and prohibit "trail associations" from any further work along these routes.[30] Always a risky scheme, the Dixie's business model had become, in many other places, an outright fraud.

The BPR and the AASHO shared Gore's suspicion of private highway associations. Shortly after the war, the BPR expressed alarm that some marked trails had "conduct[ed] a propaganda [campaign] . . . referring to their projects as national roads of importance" and in the process intimated that the federal government was about to take them over and make them national defense highways. Although the BPR did not single out the DHA, the charge must have stung a bit. AASHO officers warned against some marked-trail associations trying "to capitalize [on] the popular demand for interstate or cross-country routes by organizing trails, collecting large sums of money from our citizens and giving practically no service in return." Although the DHA never committed subscription fraud, the directors might have recognized some of their own methods, and their failings, in that description. The AASHO even passed a resolution warning citizens to investigate the claims of such groups before giving them money.[31]

Most important, the BPR and the AASHO believed that the uncoordinated plans of numerous unofficial groups complicated the already-difficult task of designating a system of federal highways. With more than 250 named routes by the mid-1920s, BPR chief Thomas MacDonald declared it "impossible" to give "official recognition" to all of them. Mac-Donald complained that many of the routes "over-lap to a considerable extent" and thus had to be "disregarded" in the process of marking the new U.S. highways.[32] In place of names, the joint board decided to use a strict, uniform numbering system. All east-west highways were to receive even numbers, while north-south routes would receive odd numbers.[33]

To federal and state officials, this system of designating interstate routes was orderly, but to proponents of named routes, it created a dilemma. Many named highways were not single-lane roads but combinations of north-south and east-west roads. The AASHO's distinction between north-south and east-west highways would break these named routes into multiple interstate highways. The Dixie Highway, with its complicated network of north-south and east-west roads, did not fit neatly into the joint board's plans. In the South, the Dixie was the longest of no fewer than a dozen marked trails that crisscrossed the region in every direction, some of them connected and some not.[34] In 1926 the DHA asserted its preference for "the preservation of Highway names such as the Dixie, Lincoln, Old Spanish Trails, etc., so that same may not be lost in the numbering system as now being carried out by the Government Highway Board," but by then the process of numbering the highways was already under way.[35]

The transformation of marked trails into numbered U.S. highways symbolized a significant transfer of power from local governments to a new centralized highway administration run by unelected state and federal bureaucrats. Unofficial highway associations like the DHA had labored for more than a decade to coordinate the work of local road commissioners to build interstate routes. The joint board's decision to number the new U.S. highway system without any regard for existing marked trails, however, threatened to erase those routes altogether. After working so hard to lay a foundation for interstate highways, marked-trail advocates would not give up without a fight.

While citizens clamored for spots on the U.S. highway system, marked-trail associations demanded special consideration for their routes. DHA officials tried first to simply retain Dixie Highway signage as an acknowledgment of their significant accomplishment. The joint board tried to distinguish between fraudulent marked-trail associations and more "reputable" groups that had "rendered distinct public service by stimulating highway improvement, maintenance, and marking." Those, the board said, would be permitted to keep their highway signs up "during their period of usefulness."[36] The DHA assumed its own "period of usefulness" to be indefinite, so directors fumed when they learned that the BPR actually would soon forbid the use of any signs other than U.S. highway numbers. Director William T. Anderson confronted Thomas MacDonald over the apparent change of plans. "It was my understanding," Anderson told MacDonald, that federal markers would merely "connect up certain routes into interstate systems and not interfere in any way with the markings" already in use. Anderson reminded MacDonald that, unlike the joint board, the DHA "has been functioning since about 1915, and was one of the pioneers in the development of road sentiment."[37] But MacDonald reminded Anderson that "a very large number of so-called trails organizations" were currently "working independently and without correlation of any kind whatever." The numerous routes had already created great confusion, MacDonald said, "to such an extent that the State highway departments had received complaints regarding the diverse markers erected on the public roads." In response, some states had already chosen to ban unauthorized markers. While MacDonald claimed that groups like the DHA were free to continue their work, they could not continue to mark the routes. He also admitted that neither the BPR nor the joint board had substituted "a uniform number" for any named route over its entire length.[38] Without either fixed numbers or the familiar red-and-white "DH" markers, the Dixie Highway would vanish from the map.

When they learned of the decision to ban unofficial highway signs, advocates of marked trails began to pursue continuous numbers for their highways instead. This was an easier task for routes such as the Coastal Highway, a single north-south highway along the Eastern Seaboard that fit neatly into the joint board's numbering system. The joint board designated most of the highway between Maine and Miami as U.S. Highway 1 but routed the 600-mile section between Richmond, Virginia, and Jacksonville, Florida, farther inland through larger towns and state capitols such as Raleigh, North Carolina, and Columbia, South Carolina. In Georgia, U.S. 1 overlapped the southeastern division of the Dixie Highway between Waycross and Jacksonville.[39]

Despite the practical reasons for the decision, residents along the southern Atlantic coastline protested the "gross discrimination" and "utter unfairness" of routing U.S. 1 inland "when this number is freely given the balance of the [Coastal] Highway between Calais, Me., and Miami, Fla."[40] Barred by new federal and state policies from maintaining even their unofficial status as Coastal Highway towns, not to mention the economic benefits of being on a marked interstate route, angry residents cried foul. S. W. Brown of Florence, South Carolina, complained to Senator Coleman L. "Coley" Blease that the new U.S. highway map "was gotten up and approved by . . . [State] Engineers, with the sole purpose in mind of Capital to Capital Highways." Brown assumed that Blease, a former governor and outspoken populist, would appreciate his misgivings about the new highway bureaucracy. The "*Columbia influence* dictated the Roads in this State," Brown charged, and he begged the senator to "see your way clear to see Mr. MacDonald of the Federal Highway department" to get U.S. 1 rerouted along the original route of the Coastal Highway.[41] Coastal Highway residents in other states also appealed to their congressmen and senators for help, despite knowing that state highway departments were handling all appeals of the joint board's initial routing decision. One North Carolinian urged his neighbors to deliberately bypass state highway officials and write directly to their congressmen, whom he believed could intervene on their behalf.[42]

The distance between citizens and federal and state highway officials, a by-product of the professionalization of highway planning, bewildered those used to dealing with elected officials. In a letter to Warren Neel in early 1926, Fred Warde, an officer in the South Atlantic Coastal Highway Association, claimed to have embarrassing evidence proving that "certain personal interests . . . [and] personal gains" were behind the joint board's selection of the route for U.S. 1. Warde claimed to have a letter "from one of the highest officials in Washington" explaining the real reason the highway had been diverted inland from the coast. This "real interesting reading matter" would not be made public, Warde promised, if the South Atlantic Coastal Highway received "justice" from the joint board. His vague threats sounded empty, but Warde's disillusionment with the new highway bureaucracy appeared genuine. He described himself and his neighbors as exemplary good-roads citizens who had worked hard with local and state officials to improve their local roads in order to serve Coastal Highway travelers. They had even supported a recent $900,000 bond issue for state highway work. Thus Warde could see no "reasonable excuse" for excluding the Southeastern Coastal Highway, so he vowed "to make a concerted protest against naming the present route Federal Highway Number One." But Neel deflected Warde's threats with a conciliatory but firm refusal to intervene, arguing that making such major changes would only invite further protest.[43]

Neel proved more accommodating to requests that he viewed as uncontroversial. The joint board's numbering system preempted the DHA's request for a uniform number for the entire Dixie Highway, but Neel supported a proposal to designate the entire eastern division as U.S. Highway 25. The plan simply called for the state highway department to remark, not reroute, an existing U.S. highway, so it gave Neel and his colleagues the opportunity to placate the DHA without upsetting anyone else. Other state leaders supported the change as well. Calling the Dixie Highway "the most important North and South Highway in the United States," the governor of Kentucky claimed that both the AASHO and the BPR already endorsed the routing change. The plan would please tourists, as well, who appreciated "the convenience of traveling on a continuous number."[44] Yet Neel's endorsement evidently did not persuade his friends at the AASHO, who selected a different route.[45] Attempts to obtain a continuous number for other marked trails also failed. Neel even intervened on behalf of the Bankhead Highway Association, only to back down when the AASHO retorted that giving the highway a continuous number would "provoke" other marked-trail associations to expect the same.[46] By mid-1926 it was clear that the U.S. highway system would absorb marked trails piecemeal.

Neel tried to steer the agency clear of controversy, but this became more difficult as the 1926 gubernatorial race got under way. With Chairman John Holder in the race, the state highway department came under intense scrutiny from both the press and his opponent, Lamartine Hardman. In this atmosphere, a local rivalry over the routing of a federal highway became front page news. In February 1926 Neel had tried to settle a bitter fight over the routing of U.S. 78 in north Georgia by persuading his state highway colleagues to propose that the AASHO mark both the Carrollton and Tallapoosa roads as alternate routes of the same highway.[47] But some citizens interpreted this effort at mediation as a strategy to win extra votes for John Holder. In a letter to the AASHO, J. J. Mangham, the mayor of a small town on the Tallapoosa route and chairman of the local good-roads club, accused the highway department of playing politics with the federal highway system. Although he defended Warren Neel as an impartial professional who "tried conscientiously to do his duty," Mangham denounced John Holder and blamed him, rather than Neel, for the decision to build an alternate U.S. 78 route through Carrollton. Calling the compromise "one of the pet political maneuvers of Mr. Holder," Mangham charged: "But for politics Carrollton would never have been heard from on this subject."[48] In a separate letter to Thomas MacDonald of the BPR, Mangham warned of the financial costs of designating alternate routes before the primary routes were even built, just to satisfy electioneering politicians like Holder. "Scattering shots," he reminded MacDonald, "rarely ever gets the game.[49]

If Mangham interpreted the alternate routing plan as an expensive and unfair electioneering scheme, both the BPR and the AASHO, who had the final say, considered it a reasonable compromise. The practice of designating alternate routes had numerous precedents in both state and federal highway work, as well as that of marked-trail associations. The Dixie Highway Association had done it on a grand scale in 1915 in order to settle numerous rivalries, and in the process it transformed a north-south highway into a vast system of interstate highways. BPR officials defended alternate routes for U.S. 78 as well. The plan not only would please residents along the Carrollton route, but it would also keep the BPR and the AASHO on good terms with their state highway department counterparts in both Georgia and Alabama, as both routes crossed the state line before merging again in eastern Alabama.[50] Even Mangham's congressman, Gordon Lee, could not convince either the BPR or the AASHO to reconsider. As a BPR official told Lee, both the Georgia and Alabama state highway departments had requested the dual routes for U.S. 78, so the decision would stand.[51] A national highway system simply mandated standardization, and local rivalries, so important in fostering citizen involvement in the Good Roads Movement a decade earlier, simply could not be taken into consideration. The endorsement of alternate federal highway routes by state and federal highway officials in the face of sustained opposition from citizens and elected officials demonstrated the power of the new highway bureaucracy.

■ ■ ■

The newfound authority of the highway bureaucracy frustrated and befuddled power brokers on the local level like Mayor Mangham and the numerous other petitioners who challenged the joint board's preliminary routing decision. Even private groups like the Dixie Highway Association balked at the AASHO's unilateral decision to disregard their wishes in the designation and marking of the new numbered routes. Although these good-roads advocates were finally getting the state and federal highway aid they had long wanted, they bristled at the consequences of empowering a group of powerful experts to administer it. In Georgia, Mangham and his supporters proved willing to eliminate one of the valuable federal highways in the state if it meant undercutting John Holder's bureaucratic politics. His grievance against the state highway department, like so many others, underscored the growing public concern about the local politics of state and federal road building.

Politics and the Rhetoric of Road Building

When the governor's race heated up in the early summer of 1926, roads were on every voter's mind. No fewer than five candidates had entered the state's Democratic primary race by June, and rumors abounded that state highway chairman John Holder would soon make it six. Two of the candidates, B. F. Mann and W. Cecil Neill, both state legislators, immediately came out as strong supporters of state highway bonds, and they, along with J. O. Wood and George H. Carswell, were outspoken critics of Holder. Dr. Lamartine G. Hardman, also a state legislator, condemned state bonds in favor of the current pay-as-you-go system, in which the state relied on gasoline taxes and auto licensing fees to generate highway revenues. This limited the state's capacity to take on ambitious highway projects but kept the state out of debt. Despite his strong feelings about highway funding, Hardman pledged early in the race to steer clear of "personal and political" issues.[52] By the time Chairman Holder entered the race later that month, however, Hardman had become his fiercest critic.

It did not take long for the campaign to turn into a two-way race between Hardman and Holder focusing almost entirely on roads. Both candidates attacked one another feverishly, but Hardman had more ammunition. He portrayed his opponent as a man who had become too powerful as head of the state highway bureaucracy, a potent allegation during the state's controversial involvement in the routing of the new U.S. highway system. Hardman capitalized on the public's fears throughout his campaign, raising sensational charges of financial corruption within

Lamartine G. Hardman, 1917. (Fred J. Bell, photographer; Lamartine Griffin Hardman Papers; courtesy of Richard B. Russell Library for Political Research and Studies, University of Georgia)

the state highway commission. He also accused Holder of conspiring with state agricultural commissioner J. J. Brown to defraud the public. Like Holder, Brown was the head of an increasingly large state bureaucracy, and his huge payroll had left him vulnerable to charges of corruption and extravagant spending.[53] For Hardman, Holder was a political opponent's dream come true.

■ ■ ■

John Holder was a powerful public figure in Georgia long before the 1926 campaign. The longtime owner and editor of the *Jackson Herald*, Holder had already served two terms as speaker of the state house of representatives when he was appointed chairman of the state highway department by Governor Thomas Hardwick in 1921. Holder's predecessor, engineering professor Charles Strahan, had been a prototypical progressive bureaucrat, but running a modern highway department required the skills of an experienced politician.[54] Holder guided the agency through a period of tremendous growth. During his tenure, he helped to transform the highway department from a small board of prison officials who acted as go-betweens to secure labor for local and federal road officials to a large team of expert engineers and civil servants who planned and built state highways. The state highway department had a negligible operating budget when Holder arrived, but by 1925 revenues from auto licensing fees

John N. Holder. (Courtesy of
Main Street News/*Jackson Herald*,
Jefferson, Georgia)

and gasoline taxes had boosted it to over $6.5 million. Although county governments still funded most roadwork in the state, applications for federal aid had to pass through the state highway department. The state highway department also selected and approved additions to the state highway system, which between 1921 and 1926 had increased by nearly 50 percent to over 7,000 miles. As head of the highway department, Holder had considerable patronage powers. By all accounts, he was adept at using them.[55]

Rumors of misconduct peaked in early 1925, when a state auditor discovered numerous accounting irregularities in the highway department's budget. Among other things, Holder had received several salary advances, which were prohibited. Holder survived Governor Clifford Walker's attempt to remove him from office, but he could not escape the mounting speculation. That summer, the state legislature opened its own investigation into charges that Holder had altered the minutes of highway board meetings, hired journalists to write favorable stories about the department's work, used department funds to pay personal expenses and looked the other way when other officials also misused departmental funds, and squandered departmental money on an expensive new car for his own use. Some witnesses also accused Holder of bribing members of the state legislature to pass the so-called midnight bill of 1922, which enhanced Holder's authority. Holder denied all the charges, but sworn testimony from colleagues William Anderson and Warren Neel suggested that there was merit to many of them. Anderson's accusations in particular prompted a rash of bitter counteraccusations from Holder. Within days, the legislature's hearings dissolved into a war of words. When it became clear that the legislature would not charge Holder with any wrongdoing, Anderson declared publicly that the legislators' probe was a "farce and [a] whitewash" designed to clear their friend Holder of all serious charges despite ample evidence against him.[56]

Holder weathered the scandal, but it tainted the reputation of the highway department. In addition to the fallout from the hearings, state highway officials faced criticism from a public dissatisfied with the pace of highway work in the state. The highway department had grown under Holder's control, but roadwork still lagged far behind. The department had added hundreds of miles to the state highway system under Holder's administration, but only a fraction of it had been completed by 1926 and hardly any of the routes had been paved.[57] Critics claimed that Georgia's roads "disgusted" tourists because they were so "indescribably bad." Holder's detractors blamed the state for the "patchwork . . . crazy-quilt" surfacing that left the highways with short stretches of good roads and very long stretches of bad roads.[58] The state's auto licensing fees and gas taxes were not enough to build continuous highways, so it still relied on counties to provide matching funds. By contrast, North Carolina had a splendid system of coordinated roads and highways thanks to $85 million in state highway bonds. In North Carolina, the state, not the counties, served as the primary "highway unit" and was responsible for coordinating and funding all state highway work. Holder's reassurances that the highway department could build a system of interconnected highways county by county fell flat. "We need more economy not more promises," one critic said, "more good roads and less loud promises."[59]

Some of Georgia's most prominent good-roads advocates argued that the only way to build better roads was to follow North Carolina's example. Holder's predecessor, Charles Strahan, believed that gas taxes were no substitute for state highway bonds. By the 1920s even Hoke Smith, the father of the state's county chain gang system and a former critic of automobile highways, was an outspoken proponent of state bonds and denounced state lawmakers who "bicker[ed] and play[ed] politics" with an issue as important as funding good roads. Speaking before a legislative commission considering a state bond issue in March 1926, Smith asked:

"Are the members of the legislature willing to make the people of Georgia pay the toll of ignorance, unpaved and unkept highways by voting to defeat the amendments to the constitution that would authorize an election for the issuance of bonds for these purposes?" Smith had retired from politics by 1926, but his advocacy of state highway bonds caused a number of highway progressives to beg him to run for governor on a bond platform. Dixie Highway Association director Clark Howell also commended Smith for speaking out for state bonds, and so did a number of other Georgians.[60]

Critics like Hoke Smith were right about the problems with the state's roads, but they misjudged public support for such a radical overhaul of the state highway department's powers. North Carolina's highway department had more in common with those in the North than its neighbors in the South. Despite dissatisfaction with progress along state highways, state highway bonds were a volatile issue in southern politics. The *Rome Tribune-Herald* printed an editorial criticizing Smith for "getting on a band wagon" for state highway bonds. "Senator Smith will have to get down to facts and show the people that he is right," the paper chided, "before he can convince them that it is necessary to borrow money with which to do something for which they already have a means of paying cash." Another local Georgia newspaper called Smith a "parasite." And while Clark Howell had praised Smith's position on state bonds, DHA director and former state highway commissioner William T. Anderson did not. "Hoke Smith came out of the departed period of Georgia politics," he said, and his appearance before the state legislature to lobby for highway bonds amounted to little more than "political advantage seeking."[61] Indeed, the gubernatorial campaign suggested that highway progressivism in Georgia was all but dead. The two candidates who launched their campaigns calling for state highway bonds in May 1926 were unheard from by late June. The two others, Carswell and Wood, devoted more time to attacking Holder's record than clarifying their own platforms.[62] Soon after Holder entered the gubernatorial contest, he and Hardman, the only two pay-as-you-go candidates, rose to the top of the primary pool.

Holder and Hardman understood the political paradox that dominated highway politics in 1926. The rural white voters who determined the outcome of the state's elections wanted state and federal highways, but they would not surrender local control in order to get them. While Holder's critics in the press and along the campaign trail denounced his past, he carefully transformed himself from an embattled bureaucrat into a viable political candidate. Although he had publicly supported state highway bonds as chairman of the highway department, Holder realized that he could not afford to do so as a candidate for political office in a state where conservative rural white voters controlled the outcome of state elections. Early in the campaign, he backtracked on bonds and declared his support for a cash-only program of road building.[63] In the face of great opposition from good-roads critics who claimed that highway bonds were the only way to build modern interconnected highways, Holder publicly vowed to give the public the roads they wanted without asking them to sacrifice control over road building to get them. "I am firmly convinced," Holder said, that by using only cash income from the gasoline taxes, Georgia "within the next seven years . . . will have . . . a most magnificent system of trunk lines and county seat to county seat roads." A few months later, candidate Holder went even further by promising voters that the entire state highway system would be paved within ten years.[64]

Holder's transition from progressive highway bureaucrat to viable political candidate was never more apparent than when he explicitly reassured voters that he did not believe in an all-powerful state highway bureaucracy. At the annual meeting of the County Commissioners Association of Georgia in Savannah in June, he cast himself as the real good-roads candidate for wanting to preserve local control. His critics who favored state highway bonds, he said, believed the state's highway administration "should be one great unit with the state highway board in full control of state highways and that no consideration should be given to the county authorities." Holder contrasted this with his own belief that "it matters not whether a road is a county, state aid or federal-aid road, it belongs to the county in which it is located." Although he believed that counties should have "jurisdiction over all roads," he did not think they should bear the burden of building them all. Holder promised once more that the state would help complete, and even pave, the state highway system, and he "confidently expected" they could do so with only a slight increase in gasoline taxes. Finally, sounding even more like a political candidate than a state official, Holder pledged that his public life had been devoted to "serv[ing] Georgia and mak[ing] her greater and better." That, he said, was his "earnest and sole desire" and would be his goal "until that solemn hour when my eyes no longer behold the light of beautiful sky, but close in eternal sleep."[65]

Lamartine Hardman was less poetic in his evaluation of Holder's accomplishments. He blasted his rival for making fantastic promises to road-deprived Georgians that he knew he could not keep. He condemned Holder for promising to build more roads for counties at the state's expense once he was in office, an act beyond the authority of a state executive though perhaps not that of the highway chairman. "How can John Holder give you something that he does not possess?" Hardman challenged.[66] He reminded an audience in the small town of Lexington, Georgia, of some of Holder's earlier, similar promises. At Holder's urging, he said, some counties had in recent years passed bonds in anticipation of receiving state aid but were still waiting for the highway department to follow through on Holder's promises. "I predict," Hardman concluded, "that they will wait several years longer."[67] Hardman began to refer to his opponent as John "Detour" Holder because as highway chairman, Holder had given Georgians more promises and more detour signs than improved highways.[68]

Despite having promised early in the race to avoid making his campaign too "personal and political," Hardman made good use of smear tactics in order to exploit the public's fear of corrupt and powerful government bureaucracy. Hardman attacked the "waste, extravagance and graft in the construction of our highways" and promised that if elected he, unlike the Holder-run highway department, would "make every dollar count."[69] In a speech to voters in Carrollton, where controversy still swirled around the routing of U.S. Highway 78, Hardman declared that the only real issue in the gubernatorial campaign was "whether the voters will permit the Brown-Holder machine to continue in office and waste the money of the state through employment of incompetent men." He continued: "I reiterate the charge that the state highway department is overburdened with expenses that should be eliminated. . . . No business man in Georgia would for a moment tolerate in his own affairs conditions that now exist in the state highway [department]."[70] Blaming Holder personally for much of the waste, he charged in another speech: "While the taxpayers are striving to meet their taxes and other obligations, Mr. Holder rides around in a four thousand dollar car at their expense." Going even further, Hardman accused his opponent of "paying salaries to several hundred men employed in the highway department to solicit support of voters for Mr. Holder." "If this isn't machine politics," Hardman asked, "tell me what is?"[71] Just weeks before the election, Hardman abandoned any constraint he had left when he and his supporters issued a statement to the press accusing Holder of being a member of the Ku Klux Klan, which by 1926 had been largely discredited by most mainstream southern politicians.[72]

Hardman reserved his most effective rhetoric to criticize John Holder's greatest political liability: his flip-flop on state highway bonds. He reminded voters that as chairman of the highway department in 1924, Holder had supported a $70 million state bond issue over the pay-as-you-go plan.[73] Along with William Anderson, then a member of the highway board as well as the Dixie Highway Association, Holder and the rest of the highway commission asked Governor Clifford Walker and the state legislature to support the bond and proposed using the gasoline tax to pay off the bond over thirty years. The bond money "could be handled practically and economically by the present minimum organization of the department," Holder and the board promised, and it would be used to pave the state highway system within seven years. The following year, in the midst of Holder's legal troubles, the highway department again asked the governor and the state legislature for a state bond issue.[74]

Under mounting pressure to clarify his position on state highway bonds, Holder went on the offensive against Hardman's allegations. Just days before the election, he told an Atlanta audience that he was "unequivocally opposed" to a state road bond issue because it was unconstitutional. Turning all of Hardman's main criticisms against him, Holder claimed that it was Hardman who was a master of "machine politics" and who changed his mind repeatedly about state highway bonds. Holder attacked his rival as part of the "Walker machine . . . the most pernicious political machine ever built in Georgia." Without mentioning his own past battles with Walker, Holder accused the outgoing governor of misusing his office to solicit votes for Hardman. As for Hardman, Holder claimed he had repeatedly changed his position on state bonds to suit his audiences. In one town, Hardman advocated a $15 million bond issue, Holder charged, in another he proposed just a "reasonable" one, and in a third he claimed he did not know where he stood on the question of state bonds. Finally, Holder accused Hardman of having close ties to the embattled commissioner of agriculture J. J. Brown and insinuated that Hardman had tried to buy Brown's support in the gubernatorial race with a $250 calf from Hardman's farm.[75]

As the election grew near, it became clear that Holder was struggling to remain afloat. He never denied that he had supported highway bonds in the past. And aside from accusing Hardman of the very misconduct he himself had been charged with before, Holder never addressed his controversial past. Although he was savvy enough to realize, unlike some of his opponents early in the gubernatorial race, that rural white voters would not tolerate state highway bonds or any other measures that increased the power of the state highway bureaucracy, his hypocrisy on the issue and the strong evidence of his misdoings as chairman of the highway commission sank his campaign. William Anderson, one of Holder's former colleagues on the highway commission, was so angry at having been implicated in some of the highway department scandals that he publicly condemned Holder and endorsed Hardman. To vote for Hardman, Anderson concluded, "was to vote for cleanliness in government," while "to vote for Holder was to vote for slap-stick, loose-jointed management."[76]

By October 1926 the race was officially between only Hardman and Holder. The other two candidates had split the anti-Holder vote in the original primary election on September 8. Carswell had done particularly well in south Georgia, his home region, weakening Hardman's lead and forcing a runoff between him and Holder. Everyone, including Hardman himself, realized the September election portended Hardman's victory in the runoff. In the statewide primary, Agricultural Commissioner J. J. Brown, whom Hardman and the other candidates had repeatedly linked to Holder, was badly defeated in his bid for reelection. "The Brown-Holder machine has been definitely smashed," Hardman confidently reported as the returns came in. The primary results demonstrated "the revolt of the people of Georgia against machine politics and in favor of a business administration in the capital," Hardman declared. "Nothing can prevent my election to the governorship." Former rivals J. O. Wood and George Carswell endorsed Hardman in the runoff, virtually guaranteeing him the rural Georgia votes that had cost him a more-decisive victory over Holder in September.[77]

On October 6, Lamartine G. Hardman won the governor's office in a landslide. He received 282 county unit votes to Holder's 132, a margin of more than two to one. Yet according to the popular vote, Hardman won by a closer margin of less than 15 percent, suggesting that Holder had more support among the state's urban voters than rural voters. Roads not only took center stage in the governor's race, then, but Holder's past record as an opponent of the conservative pay-as-you-go plan clearly resonated with Georgia's dominant rural voters.[78] So, too, did the prospect of mitigating the power of the highway bureaucracy. As the *Atlanta Constitution* exclaimed in a postelection editorial, Hardman's victory "reflected the determination of the people to apply business to government" and declared an end to the "petty politics" of the past.[79] In a final insult to Holder, Hardman declined to reappoint him as highway chairman in 1927. In his place, Hardman appointed Holder's predecessor, Charles Strahan, a move that embodied the determination to turn back the clock on the state's highway work.[80]

Meanwhile, the pressure on state highway departments only increased. A month after the Georgia election, the AASHO met to vote on the numerous route changes and additions proposed by state highway commissions. Led by Warren Neel, Georgia highway officials arrived with only four requests, having either settled or refused the numerous other requests they received. The other states requested additions as well, and together those that were approved increased the total mileage of the new U.S. highway system to 96,626 miles, nearly double the mileage proposed by the joint board just a year earlier. The states voted to approve the additions on November 11, 1926.[81]

It was no accident that the demise of the Dixie Highway Association dovetailed with the designation of the federal highway system. By April 1926, most states had ratified the proposed system of numbered federal highways, and by the following spring, most of the changes and additions requested by the states had been approved.[82] On April 22, 1927, the executive committee of the DHA met one last time to vote on a resolution to end the active operation of the group. Agreeing that "the purpose for which the Dixie Highway Association was organized has been accomplished" and that the DHA no longer had the funding to pay its basic expenses, five officers voted unanimously to end the association's activities and maintain it in name only "for any future needs."[83]

■ ■ ■

After politicians moved on to new debates, Warren Neel, the highway expert who had always been more comfortable building roads than playing politics, insisted on the recognition of the unique merits of the Dixie Highway. Through some misunderstanding, the AASHO had failed to designate the Dixie Highway route between Rome and Cartersville as part of the U.S. highway system. In his detailed appeal to the BPR on behalf of supporters of the route, Neel summed up more than a decade of work on the Dixie Highway. Not only did twenty or thirty miles of the Rome route pass through Chickamauga Park, "which, as you know, is of great historical value," it also "is already heavily traveled and the best route at this time from Chattanooga to Atlanta and is rapidly being paved." He declared that "practically all of the travel coming in to Georgia from the Middle West comes in thru Chattanooga and follows this route," while traffic from the East entered Georgia on the Dalton branch of the Dixie. Comparing the Dixie Highway to other marked trails, Neel insisted that both routes between Chattanooga and Atlanta "have considerably more travel than have the two branches of the Bankhead Highway . . . [which] were approved for alternate marking." Lest his request again get lost in a sea of others, Neel begged "that this request . . . take precedence over others."[84]

Broken into countless pieces, many sections of the Dixie Highway from Lake Michigan to Miami Beach became U.S. highways. In the South, these portions of the federal highway system represented the culmination of years of work. Though bittersweet for DHA officials who had hoped to keep Dixie Highway markers along the original route, it was a triumph for progressive good-roads advocates like Neel, who recognized that the Dixie Highway had succeeded where so many others had failed.

But as the debates over chain gangs, state funding, and the power of the highway bureaucracy foretold, the full measure of that success would take another thirty years to realize. At the federal level, concerns over extending federal highway aid in the mid-1920s echoed those in the South. In his 1925 budget proposal, the parsimonious President Calvin Coolidge asked Congress to limit federal highway aid, arguing that it was a drain on the U.S. Treasury and "detrimental" to federal and state governments alike. Coolidge believed that sharing the costs of the federal-aid program "unduly enlarged" the federal government and "impaired" the efficiency of state governments by ceding to federal authorities responsibilities that belonged to the states. It was, he concluded, an "insidious practice which sugar-coats the dose of Federal intrusion." Many congressmen agreed with him.[85] Congress temporarily reduced federal aid appropriations in 1923 from the original $75 million to $50 million before restoring it two years later. Congress did not increase appropriations again until the 1930s, when federal relief projects required it, but by 1936 it had settled at

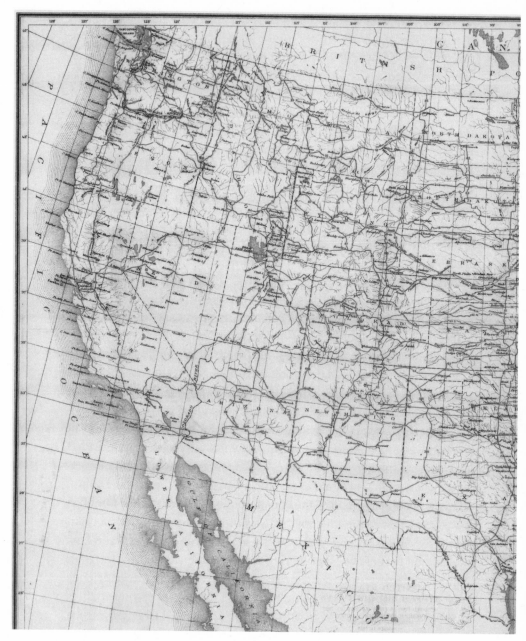

U.S. highway system, designated November 1926.

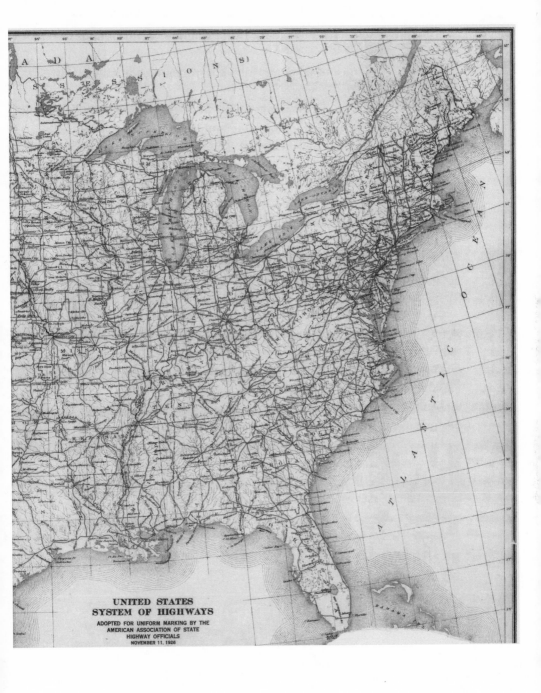

**UNITED STATES
SYSTEM OF HIGHWAYS**

ADOPTED FOR UNIFORM MARKING BY THE
AMERICAN ASSOCIATION OF STATE
HIGHWAY OFFICIALS
NOVEMBER 11, 1926

$125 million, where it remained until World War II. In the meantime, the federal-aid system expanded slowly, despite the designation of the new U.S. highway system, and states and local governments continued to pay for the lion's share of roadwork.[86]

In the 1930s and 1940s, the highway bureaucracy continued to grow, albeit unevenly. Even after Congress implemented new federal gas taxes in 1932, federal highway aid languished. There was little demand for additional revenues when states could not afford to match them. The onset of the Great Depression caused state and local revenues to plummet, especially in the South, where they were never sufficient to begin with. During the Depression, many states diverted their own gas-tax revenues to nonhighway expenses, but good-roads advocates in the southern states were slower than most others in stopping the practice. Not until the late 1940s and 1950s did some southern states finally pass "linkage" laws earmarking gas-tax revenues for highway construction only.[87]

In Georgia, as in the rest of the South, the backlash against the expansion of the highway bureaucracy that culminated in the tumultuous gubernatorial election of 1926 had a lasting effect. Federal aid expenditures in Georgia declined starting in the 1920s, and both local road and state highway construction in the state stagnated for a quarter of a century thereafter.[88] In many ways, the race between Hardman and Holder placed the massive success of the Dixie Highway campaign up for a popular vote, and it lost.

Conclusion

The Georgia scenario would soon play out across the South. Despite the desperate need for better roads and a consensus in favor of building them, voters recoiled at the power and the cost of the state and federal institutions necessary to implement actual road construction. By the final years of the 1920s, southern politicians created careers out of stoking populist outrage against taxes and government. In this new turn of the political tide, state highway departments became favored targets for angry screeds against bloated government and state corruption.

Such popular outrage flared up throughout the rural South, but nowhere was it more apparent than in Georgia. In the Georgia elections of 1926, Agriculture Commissioner J. J. Brown, with whom John Holder's name had been so often linked in scandal, lost his bid for a sixth term in office. Brown's successor as commissioner of agriculture was a newcomer named Eugene Talmadge. Soon to become one of Georgia's most famous demagogues, Talmadge cultivated rural white support with the same grassroots populism that doomed Brown and Holder. Commissioner, and later Governor, Talmadge rose to power by harnessing the grassroots voting power of Georgia's struggling rural farmers. He spun a political message that was a powerful mix of antitax, small-government philosophy; white supremacy; and cult of personality. "The poor dirt farmer," he was fond of saying, "ain't got but three friends on this earth: God Almighty, Sears Roebuck, and Gene Talmadge."[1]

The state highway department remained Talmadge's bête noire throughout his career. Although the state highway department had emerged from the Good Roads Movement, a largely agrarian, grassroots social movement, Talmadge rhetorically transformed it into what one of his biographers has called a "a monstrous, uncontrollable appendage of state government, and . . . another populistic enemy [for Talmadge] to battle." Talmadge's message seemed particularly salient during the economic crisis of the 1930s. By the onset of the Depression, farmers who had become dependent on their cars and trucks found themselves increasingly hard-pressed to pay the tag fees that supported road maintenance. Talmadge played upon popular discontent with auto fees to fuel his campaign against the state highway department.[2]

During Talmadge's first gubernatorial run in 1932, he met a man who had borrowed his neighbor's tag to haul a load of firewood into town to sell. With tag fees ranging from $8.00 to $25.00, the man said he simply could not afford to buy his own. "Gene," the man complained, "I gotta git a three-dollar tag for this truck that cost me five dollars. . . . If they would lower the cost of tags we could buy more gas to offset the loss [in highway revenue]." Struck by the prospect of relying more on gasoline taxes, which everyone who passed through the state paid, instead of higher tag fees, which strained only his constituents, Talmadge found a way to use the Great Depression to further fuel his crusade against big government and the highway department. Throughout his successful campaign, Talmadge seized upon the idea to limit the state's auto licensing fee to a flat rate of $3.00 as a way of relieving farmers of "the heavy burden of taxation" that compounded their problems during the economic crisis. Fiddlin' John Carson wrote a campaign song for Talmadge that skewered high automobile tag fees:

I've gotta Eugene dog, I gotta Eugene cat,
I'm a Talmadge man from ma shoes to ma hat.
Farmer in the cawnfield hollerin' whoa gee haw,
Kain't put no thirty-dollar tag on a three-dollar car.[3]

Talmadge attributed his overwhelming victory in 1932 to the popularity of his three-dollar tag fee position, and his win threatened to eviscerate the state highway department. The highway department made nearly $9 million from state tag fees in 1931 alone, a figure that would dissolve after the implementation of Talmadge's three-dollar tag proposal. State legislators hesitated to slash the budget of such a crucial agency, but the populist sentiment spoke powerfully enough for Talmadge to ram the measure through. Talmadge also refused to sign off on new sales taxes to generate much-needed state revenues, and he tried to seize control of several powerful state agencies, including the highway department. Though the state legislature restrained some of his most egregious abuses of executive power, Talmadge remained wildly popular among the state's rural voters. His victory had opened up a powerful vein of popular discontent, one suspicious of state agencies and the taxes that sustained them. When the three-dollar flat fee initiative passed, voters greeted the news with cheers, and Fiddlin' John Carson marked the occasion with a new song: "Got chickens in the coop, eggs in the bag / Eugene's got us our three-dollar tag"[4]

Talmadge's war against the state highway department continued after the tag battle, and his message spread to other southern states along the Dixie Highway. In 1936 South Carolina governor Olin Johnston attempted to take over the state highway department through executive reorganization. When the federal Bureau of Public Roads rebuked Johnston for his actions, the power struggle merely served to embitter rural constituencies even further.[5] The BPR also intervened in Georgia in 1935, temporarily cutting off federal aid funding over concerns about Talmadge's meddling with highway commission appointments. And only intervention by progressive state and federal officials in the 1930s and 1940s curbed the practice of diverting gasoline taxes from state highway department budgets.[6] Manipulated by skilled demagogues like Talmadge and Johnston, these battles suggested to rural southerners that their troubles arose in distant bureaucracies, not in reactionary politics closer to home. Poor, struggling farmers increasingly placed their faith in their popular governors and focused their contempt on unelected functionaries and their powerful state and federal agencies.

That these conflicts emerged during the height of New Deal–era federalization was no coincidence. Although many rural southerners benefited from federal relief programs, their allegiance to President Franklin Roosevelt and New Deal programs was shaky at best. Some were drawn to critics of New Dealism, such as Louisiana firebrand Huey Long, whose populist appeal as governor of and then U.S. senator from Louisiana rivaled only perhaps that of Eugene Talmadge. While detractors said that Long's famed "Share Our Wealth" program smacked of socialism, it amounted to little more than a popular but untenable critique of Roosevelt's failing New Deal policies in the South.[7] As historian Shane Hamilton has shown, many rural Americans spurned New Deal liberalism and bureaucratic federal regulations in favor of free-market philosophies that shaped the future of commercial agriculture and the rise of postwar conservatism alike.[8] In the context of such critiques of federal power, the backlash against the highway bureaucracy in southern states makes all the more sense.

■ ■ ■

Political struggles over highways, the reach and limit of state power, and the high cost of roads flared up repeatedly from the 1930s through the 1950s. The collective impact of such struggles effectively derailed—and ultimately eclipsed—the pioneering developments in road building and highway policy making of the 1910s and 1920s. The project that began along the Dixie Highway required more money and centralized management than any previous domestic public-works project. Throughout the Dixie's short life, citizens, lawmakers, and expert professionals grappled with the best way to imagine and execute a road-building project on an unheard-of scale. In the end, the project became mired in a political dilemma as pressing to the good-roads campaign as it is today: how can a government sustain a massive internal-development project when there are limits to the public's willingness to pay for it?

It would be two decades before the nation would be ready to embrace the idea of a new federal interstate highway system. How could an idea that was rejected in the 1920s and 1930s emerge again on such an enormous and triumphant scale as the Eisenhower Interstate System did in the 1950s? It happened in part because the Dixie Highway never actually disappeared. Carl Fisher's ideas, hatched in 1914, opened doors that could not be shut, even though politicians vigorously tried to do so. Although the country remained loath to embrace all that interstate highways entailed in the 1920s, the project set in motion new ways of conceiving of roads that fundamentally altered the national discourse about transportation and investment.

The politics of national highways, interstate travel, state power, and federal investment shaped debate, albeit in new forms, throughout the 1930s and 1940s. During the 1930s, economic stimulus replaced public demand as the primary motivation for federal highway aid, obscuring some of the ongoing political infighting. Stricter highway standards from the Bureau of Public Roads and the increasing demands of automobile traffic put an enormous strain on the capacities of Depression-era state and local governments. The 1934 federal aid bill threatened to penalize states that diverted gas-tax revenues to their general funds, but the shortfalls continued, creating an enormous void for President Roosevelt and the New Deal to fill.[9] Over a third of federal work-relief jobs were on road projects, many of which were not even under the control of the federal Bureau of Public Roads. Many, moreover, were in urban areas. These projects had more to do with relieving the unemployment crisis than with road building, so they concealed the bureau behind New Deal policy making, undermined the established program of federal aid, and limited the reach of federal and state highway agencies.[10]

Although work-relief jobs did not necessarily ensure much new mileage, the New Deal enabled the BPR to improve existing roads and build strategically placed new ones that remain today among the most popular tourist routes in the nation. During the 1930s, the BPR built over 1,500 miles of good roads in and around the nation's national parks and monuments, including impressive paved highways such as the Skyline Drive in Virginia and the Blue Ridge Parkway along the crest of the Appalachian Mountains in North Carolina. Moreover, the Civilian Conservation Corps put many southerners to work improving roads and bridges in some of the most isolated parts of the region, while other work-relief employees improved city routes to serve the growing urban population. In spite of vast increases in federal spending, basic highway policy remained unaltered in the 1930s. The cooperative model of road building established during the previous two decades actually flourished under the direction of the BPR.[11]

Although Roosevelt's New Deal programs channeled unprecedented federal funds to highway work—$2.8 billion between 1934 and 1937, for example—Roosevelt resisted pressure to permanently and systematically expand federal highway funding. As historian Mark Rose has argued, Roosevelt was uncertain about the economic impact of highway work, and he preferred to keep federal spending within pre-Depression limits and channel more money into other programs that he felt created more permanent jobs. Resolving the long-term economic crisis, Roosevelt believed, was more about jobs than roads.[12]

In the late 1930s and 1940s, as traffic congestion grew worse, the state continued to dance around the same debates over the size of government and the structure and use of taxes. Roosevelt's high-profile highway-parkway projects made for good New Deal publicity, but they did not address either the need for more high-traffic routes or the political divisions that preempted their construction. Some states and private companies began building toll roads to meet the popular demand for highways. The BPR, the AASHO, and the automobile industry joined forces to protest the expansion of toll roads, proposing instead that the federal government build public-access highways linking the centers of the nation's major cities.[13]

In the wake of World War II, motorists, commercial truck operators, and state highway engineers joined them. Together they formed a new highway lobbying campaign that resembled the Progressive Era Good Roads Movement. In this new context, they repeated the same debates that had emerged during the First World War. Commercial and military needs, more linked after World War II than ever before, demanded renewed attention to the nation's most heavily traveled routes. Only toll-free express highways, they argued, could meet the country's postwar needs.[14]

By the end of the Second World War, the new highway lobby, with presidential leadership, finally convinced Congress to increase funding for federal highways. In 1956 President Eisenhower signed the Federal-Aid Highway Act, the largest and most comprehensive federal highway plan in the nation's history. The bill provided $25 billion for over 40,000 miles of federal interstate highways—a "quantum leap," in the words of one historian—in federal highway funding. The move signaled that the federal government had finally taken primary responsibility for meeting the demands of an automobile-dependent public.[15]

Three decades after the Dixie Highway experiment ended, federal support for highways began to transform domestic transportation. In Georgia, the federal government's assumption of responsibility for most main highways freed state and local officials to focus exclusively on their own roads for the first time. In 1958 a local newspaper praised the state's Rural Roads Program, a $100 million initiative to improve at least three country roads in each of the state's 159 counties. By 1965 Georgia had over 96,000 miles of roads, a third of which had been paved, which was the same as the national ratio. Today, the state has some 118,000 miles of roads, including over 1,200 miles of federal interstate highways, 18,000 miles of state highways, 84,000 miles of county roads, and 14,000 miles of city streets.[16]

Yet considering the vast changes in the administration of roads and highways, these numbers have changed very little since the 1920s. The most significant increases are not in total mileage but in the proportion of state and federal highways to county roads. Counties still maintain the vast majority of the state's roads, but the state and federal governments maintain thousands of miles of the most heavily traveled routes. These do not include the old path of the Dixie Highway in Georgia, which has been replaced by Interstate 75. Indeed, very little of the original roadbed survives, even as local roads. Except for a short stretch south of Atlanta and another a few miles south of Macon, the old Dixie Highway is long gone.

The process of road building has been as fraught with problems in recent years as it was in the first half of the twentieth century. Today, over 300 million Americans own 250 million automobiles. Georgians own over 8 million automobiles, putting the state's per capita ratio at one car for every 1.25 people, exactly the same as the national ratio. Americans travel about 3 trillion miles a year on the nation's nearly 5 million miles of roads, including nearly a million miles of federal-aid highways that are not included in the federal interstate highway system. In contrast to the late 1910s and early 1920s, counties are responsible for less than half of federal-aid routes, while states maintain about 60 percent of them. As mileage has increased, then, so has automobile traffic, so the demand for roads has kept a steady pace with the progress of road building. The only difference is that federal and, especially, state highway departments now shoulder the enormous burden once borne by county governments.[17]

Overwhelmed by their responsibilities to a car-centered American public, some states have recently begun turning back to the private sector for help. These new initiatives depend upon profit, not cooperation. Unlike the public-private approach to building the Lincoln Highway or the Dixie Highway, this new partnership bypasses the administrative structure created through those old highway initiatives. In 2005 in deficit-ridden California, for instance, Governor Arnold Schwarzenegger approved a plan allowing private investors to build and maintain toll roads. "Californians can't get from place to place on little fairy wings," he said. "We are a car-centered state. We need roads." In other states, existing toll roads are being privatized. In 2005 the Spanish Cintra-Macquarie consortium paid nearly $2 billion for a ninety-nine-year lease on Chicago's eight-mile Skyway and twice that much to operate the Indiana Toll Road for seventy-five years. Harris County, Texas, is also considering selling or leasing its county toll system.[18] Elsewhere in Texas, corporations are building the largest private highway system in the nation. The Trans-

Texas Corridor, a $185 billion project, will incorporate thousands of miles of roads built with tax dollars. Nearly two dozen other states have also announced plans to privatize public roads within their borders. Georgia does not seem to be far behind this trend. Although the state relies more heavily on federal aid and bonds than most other states, the legislature is wary of raising the gasoline tax to pay for the roads that Georgians need.[19]

Debates over highway initiatives today mirror those of nearly a century ago. Voters in states across the country regularly reject proposals for pricey new transportation projects ranging from interstate highway bypasses to public transportation initiatives, even as they complain about the crisis over the nation's crumbling, increasingly insufficient highway infrastructure. Once again, these debates break down along regional lines that underscore the intensely local nature of transportation politics. Voters in cities do not want to pay for highways that serve suburban commuters and rural people, while those voters in turn refuse to finance public transportation alternatives that they will likely never use themselves.[20] Even in Congress, debates over continuing federal highway funding break down in ways that reflect sectional and political divisions and often lead to compromise legislation that pleases no one.[21] Taken collectively, these examples serve as a reminder that the ongoing struggle to manage our transportation infrastructure is an expensive and controversial national crisis, and it always has been.

Notes

Abbreviations

AC	*Atlanta Constitution*
BPR	Records of the Bureau of Public Roads, Record Group 30, National Archives and Records Administration, Washington, D.C.
DH	*Dixie Highway*
DHAMB	Dixie Highway Association Minutes Book, Records of the Chattanooga Automobile Club/Dixie Highway Association, Chattanooga–Hamilton County Bicentennial Library, Chattanooga, Tenn.
ENR	*Engineering News-Record*
GDAH	Georgia Department of Archives and History, Atlanta, Ga.
HSP	Hoke Smith Papers, Richard B. Russell Library for Political Research and Studies, University of Georgia, Athens, Ga.
LE	*Leader-Enterprise* (Fitzgerald, Ga.)
LHC	Lamartine G. Hardman Collection, Richard B. Russell Library for Political Research and Studies, University of Georgia Libraries, Athens, Ga.
NYT	*New York Times*
RFRC	Rome-Floyd Records Center, Rome, Ga.
RTH	*Rome Tribune-Herald*
TRP	Governor Tom C. Rye Papers, Tennessee State Library and Archives, Nashville, Tenn.

Introduction

1. Stager and Carver, *Looking beyond the Dixie Highway*; Cox, *Dreaming of Dixie*, 144–46.

2. On early twentieth-century roads, see Holley, *The Highway Revolution*, and Preston, *Dirt Roads to Dixie*. On the Lincoln Highway, see Hokanson, *The Lincoln Highway*. Most other books about the Lincoln Highway are either travel guides or coffee-table books.

3. Many of the most prominent histories of the New South ignore roads altogether. See Ayers, *The Promise of the New South*; Woodward, *Origins of the New South*; and Tindall, *The Emergence of the New South*. Historians who do mention the impact of roads on politics typically dispense with the topic in a single chapter or subsection. See, for example, Kirby, *Rural Worlds Lost*; Ownby, *American Dreams in Mississippi*; and Simon, *A Fabric of Defeat*. On southern Progressivism, see Link, *The Paradox of Southern Progressivism*.

4. Hilles, "The Good Roads Movement in the United States, 1880–1916." Seely, *Building the American Highway System*. Gutfreund, *Twentieth-Century Sprawl*. Hugill, "Good Roads and the Automobile in the United States, 1880–1929." McKown, "Roads and Reform." Lesseig, "'Out of the Mud.'" Sutter, "Paved with Good Intentions." Keith, "Lift Tennessee out of the Mud."

5. Preston, *Dirt Roads to Dixie*, 1, 5.

6. A recent example that focuses on modern roads and highways is Gutfreund's *Twentieth-Century Sprawl*, an excellent study of the impact of automobiles and highways on communities. Mark Rose's classic work, *Interstate*, offers a very good overview of midcentury highway construction from the perspective of an urban historian. Seely's *Building the American Highway System* devotes just one chapter to road building and focuses instead on the experts behind the nation's evolving highway policies. Jakle and Sculle's *Motoring* focuses on the motoring experience in American culture and offers only a broad, thin summary of road building. Among the best studies of the early auto industry is Flink, *America Adopts the Automobile, 1895–1910*.

7. Hall and others, *Like a Family*, xi, 46–47. Carlton, *Mill and Town in South Carolina, 1880–1920*, 46–49. Flamming, *Creating the Modern South*, 19.

8. Preston, *Dirt Roads to Dixie*, 158. USDA Office of Public Road Inquiry, "The Railroad and the Wagon Road."

9. Carlton, *Mill and Town in South Carolina*, 21–25. Flamming, *Creating the Modern South*, 22.

10. Hamilton, *Trucking Country*, 44–47.

11. Fuller, "The South and the Rural Free Delivery of Mail," 506.

12. Stager and Carver, *Looking beyond the Dixie Highway*.

13. Flamming, *Creating the Modern South*, 178–79. Flamming notes that the bedspread industry began earlier but expanded beyond local markets due to the Dixie Highway tourist trade and better access to distant markets.

14. Link, *Paradox of Southern Progressivism*.

15. Okrent, *Last Call*, 107. Engelmann, *Intemperance*, 34. Draper and others, "Michigan's 'Great Booze Rush' and Its Suppression by State and Federal Action," 85–90. Even after federal Prohibition ended, the lucrative moonshining business in the South was utterly dependent upon good "trippers" who could outrun the law in fast cars along bumpy back roads. Some of the most successful trippers pioneered the sport of stock-car racing, which attracted an international following after the creation of the National Association for Stock Car Auto Racing (NASCAR) in 1948. On the origins of stock-car racing, see Daniel, *Lost Revolutions*, 91–120; and Pierce, *Real*

NASCAR. Randall L. Hall downplays the rural moonshining origins of stock-car racing but concedes that trippers contributed to the sport. See Hall, "Before NASCAR."

16. This is the essence of C. Vann Woodward's famous assertion that Progressivism was for whites only. See *Origins of the New South*, 369–95. On prison labor, see Lichtenstein, "Good Roads and Chain Gangs in the Progressive South."

17. The Golden Gate Bridge was not completed until the 1930s, but it was planned in the late 1910s.

Chapter 1

1. "Bad Roads Did It," *Southern Good Roads* 1, no. 4 (April 1910): 18. The poem has three more verses.

2. Address by Rev. Joseph Hayes Chandler to the Chicago good-roads convention, 1912. Read into the record of the 62nd Cong., 2nd sess., *Congressional Record*, 599. Chandler attributes the apocryphal story to Abraham Lincoln. A slightly different version of this story is cited in Holley, *The Highway Revolution*, 4.

3. For an overview of early twentieth-century roads, see Holley, *The Highway Revolution*. On roads in the South during this period, see Preston, *Dirt Roads to Dixie*.

4. Hugill, "Good Roads and the Automobile," 332. Seven more had established state highway departments by 1905, but most other southern states did not follow until the 1910s. Seely, *Building the American Highway System*, 24.

5. On the history of Native American trading paths in the South, see Hudson, *Creek Paths and Federal Roads*.

6. Postel, *The Populist Vision*, 146–47.

7. Hulbert, *The Future of Road-Making in America*, 16.

8. Flink, *America Adopts the Automobile*, 145–50. Preston, *Dirt Roads to Dixie*, 163–64.

9. Daniel Okrent surveys many of these opportunistic alliances in *Last Call*. On disfranchisement, see Perman, *Struggle for Mastery*.

10. U.S. Department of Agriculture, Office of Public Roads, "Mileage and Cost of Public Roads in the United States in 1909," 40–120. Other southern states were in line with the figures in Georgia, except for Mississippi, which had improved less than 1 percent of its public roads. Georgia was also close to the national average, making it a better state for a case study than exceptions such as Mississippi. These figures do not count the hundreds, perhaps thousands, of smaller roads linking farms to the main public roads of the county. No one had ever calculated the actual total mileage in the United States. See Seely, *Building the American Highway System*, 12.

11. "Where the South Stands," *Southern Good Roads* 1, no. 6 (June 1910): 20.

12. Roy Stone, "Address at Manassas, Va.," September 1895, Box 2, Vol. 4, Central Correspondence: Letters Sent by the Bureau, 1893–1904, BPR. Stone also cited instances of corruption and greed, including the use of the county road fund as a "pension fund" for the road commissioner's friends. Georgia counted the value of forced labor toward the state's road expenditures; in 1910, the amount was $450,000, or nearly 20 percent

of the state's total road expenditures. See Geological Survey of Georgia, "A Second Report on the Public Roads of Georgia," 1.

13. Road Supervisors Book for Chattooga County (1886), DOC 6744, RG127/SG12/S14, Box 18, GDAH.

14. Sheffield, "Improvement of the Road System of Georgia," 24.

15. Seely, *Building the American Highway System*, 12. Sheffield, "Improvement of the Road System of Georgia," 8.

16. U.S. Department of Transportation, Federal Highways Administration, *America's Highways*, 84. Hulbert, *The Future of Road-Making in America*, 16. If these figures seem high, consider that agriculture experts in North Carolina had estimated that state's losses alone to be $10 million in 1902. See "Economy of Good Roads," *Proceedings of the North Carolina Good Roads Convention, February 1902*, USDA Bulletin No. 34, 46.

17. Some figures collected by the USDA's Office of Public Roads are cited in congressional debates. See Senator Candler, "Road Improvements," *Congressional Record* 48, pt. 1 (December 19, 1911): 525.

18. Holley, *The Highway Revolution*, 4–6. Flink, *America Adopts the Automobile*, 101. Hulbert, *The Future of Road-Making in America*, 16. Preston, *Dirt Roads to Dixie*, 158.

19. Rigdon, *Georgia's County Unit System*. Kytle and Mackay, *Who Runs Georgia?*, 9–13. Bartley, *The Creation of Modern Georgia*, 161–66. The system did not actually become law until 1917 with the Neill Primary Act. The only time the county unit system resulted in a major upset of a state-level election was Eugene Talmadge's famous gubernatorial victory in November 1946. Except for a brief period in 1908, the county unit system remained legal until 1962.

20. Speech of Honorable Hugh M. Dorsey of Atlanta, July 28, 1914, Series III, Box 2, Folder 14, HSP. Dorsey was campaigning for former governor Joseph M. Brown, who ran unsuccessfully for the U.S. Senate in 1914. Dorsey was an Atlantan himself, but he was also a conservative politician who needed the county unit system in order to be elected. Governor Hoke Smith temporarily abolished the county unit system in 1908, but Brown (his successor in the governor's mansion) reinstated it. See Grantham, *Hoke Smith*, 172–93.

21. Southerland and Brown, *The Federal Road through Georgia, the Creek Nation, and Alabama*, 8–9.

22. Hilles, "The Good Roads Movement," 4–12.

23. Dearing, *American Highway Policy*, 31–34. Carter, *When Railroads Were New*, 4. On the Federal Road, see Southerland and Brown, *The Federal Road through Georgia, the Creek Nation, and Alabama*; and Hudson, *Creek Paths and Federal Roads*.

24. Stover, *American Railroads*, 10–34, 61–86; for railroad mileage, see page 205. Stover argues that private capital funded most early railroads (before 1850) and downplays the impact and ultimate cost of federal land grants to railroads in the second half of the century. Gordon, *Passage to Union*, 151–52. Dearing, *American Highway Policy*, 36–39.

25. Balogh, *A Government Out of Sight*. Stover, *American Railroads*, 84.

26. Hilles, "The Good Roads Movement," vi (Figure 1).

27. Postel, *The Populist Vision*, 146. For more on railroad abuses, see Stover, *American Railroads*, 96–132.

28. Postel, *The Populist Vision*, 146–50. Stover, *American Railroads*, 96–132.

29. Gordon, *Passage to Union*, 201–5. Dearing, *American Highway Policy*, 46.

30. Hilles, "The Good Roads Movement," 24–48. *Statistics of the Population of the United States at the Tenth Census, 1880*, xxix–xxx. Potter, *The Gospel of Good Roads*, 5–6, 52–59.

31. Preston, *Dirt Roads to Dixie*, 29–30.

32. Seely, *Building the American Highway System*, 12–18. Hilles, "The Good Roads Movement," 49–53.

33. Roy Stone to J. Sterling Morton, December 23, 1893, Box 1, Volume 1, Central Correspondence, BPR. Stone actively lobbied railroad officials to get them to offer discounted services. See, for example, Stone to C. J. Ives, May 8, 1894, and Stone to S. W. Fordyce, May 8, 1894, Box 1, Volume 1, Central Correspondence, BPR. Sometimes Stone had difficulty convincing railroad officials to cooperate. See Stone to Stuyvesant Fish, President of the Illinois Central Railroad, May 7, 1894, Box 1, Volume 1, Central Correspondence, BPR.

34. Seely, *Building the American Highway System*, 18–19. Goddard, *Getting There*, 52. Seely notes that Dodge sometimes pushed the limits of legal partnerships with the railroads.

35. Woodward, *Tom Watson*, 245–46. Postel, *The Populist Vision*, 143–46. Hilles, "The Good Roads Movement in the United States," 174–75. Watson was not the first to propose Rural Free Delivery. Illinois editor John M. Stahl began a public campaign for rural mail delivery in 1879, and Postmaster General John Wanamaker convinced a reluctant Congress to let him experiment with it shortly before Watson's proposal. See Leach, *Land of Desire*, 182–83.

36. The nation's rural population was two-thirds of the total population, down slightly from about three-fourths in 1880 and 1890. See *Twelfth Census of the United States, Taken in the Year 1900*, part 1, *Population*, xviii and lxxxii.

37. Kirby, *Rural Worlds Lost*, 117–18. Leach, *Land of Desire*, 183–85.

38. Preston, *Dirt Roads to Dixie*, 19–20, 29–30.

39. Pope is quoted in Flink, *America Adopts the Automobile*, 202–3.

40. Ibid., 12–25, 275, 308–9. Flink, *The Automobile Age*, 24–26. *Thirteenth Census of the United States (1910): Abstract of the Census*, 504–5.

41. Flink, *The Automobile Age*, 34–39. Flink notes that not until 1912 did the price of even the popular Ford Model T drop below the average annual wage in the United States.

42. Berger, *The Devil Wagon in God's Country*, 13–26.

43. "Calls AAA Roads 'Peacock Lanes,'" *Automobile Topics* 33 (February 1914): 178. Goddard, *Getting There*, 61. On peacock alleys in hotels, see Host and Portmann, *Early Chicago Hotels*.

44. Flink, *America Adopts the Automobile*, 64–66. Flink notes that farmers' hostility to motorists was short-lived. Hilles, "The Good Roads Movement," 86–87. Berger, *The Devil Wagon in God's Country*, 14–28.

45. McGerr, *A Fierce Discontent*, 107–11, 132–36, 147–81. The Keating-Owens Act was later declared unconstitutional, and there was no further federal child-labor legislation until Franklin D. Roosevelt's Fair Labor Standards Act of 1938. On tariffs, see Flink, *The Automobile Age*, 44.

46. Flink, *America Adopts the Automobile*, 301–6; The panic of 1907, which disrupted credit arrangements for even the most stable automakers, also played a role in driving many vulnerable automakers out of business. See Madsen, *Deal Maker*, 108–9.

47. Ford, *My Life and Work*, 21–32. Watts, *The People's Tycoon*, 3–41. Flink, *America Adopts the Automobile*, 268–78; Flink, *The Automobile Age*, 37.

48. Flink, *The Automobile Age*, 37–38. Ford sold two Model T cars: the runabout and the larger and slightly more expensive touring car. In 1910 the touring car cost $780, and in 1916 it was $360.

49. Flink, *America Adopts the Automobile*, 278–92.

50. Ibid., 285.

51. Rae, *The American Automobile*, 50–51. O'Reilly, *The Goodyear Story*, 18–21.

52. Foster, *Castles in the Sand*, 12–30. Herlihy, *Bicycle*, 251.

53. Jane Fisher, *Fabulous Hoosier*, 49–51. Foster, *Castles in the Sand*, 37–40. The ad is quoted in Jerry Fisher, *The Pacesetter*, 37.

54. Foster, *Castles in the Sand*, 45–59.

55. Jerry Fisher, *The Pacesetter*, 39–59. Foster, *Castles in the Sand*, 70–95. Jane Fisher, *Fabulous Hoosier*, 47–54.

56. "Sketches Show Caliber of Marion County's Progressive Nominees," *Indianapolis Star*, September 15, 1912, 9. "Fisher's Salary to Go to Roads," *Indianapolis Star*, October 24, 1912, 3. Foster, *Castles in the Sand*, 102–2.

57. "Fisher's Salary to Go to Roads." Fisher also worked through groups such as the Hoosier Motor Club. See "Good Roads Held Absorbing Topic," *Indianapolis Star*, November 10, 1912, 20. Jane Fisher, *Fabulous Hoosier*, 79.

58. Rose, *Interstate*, 2.

59. Flink, *America Adopts the Automobile*, 75–86. Preston, *Dirt Roads to Dixie*, 163–64.

60. Quotations in Preston, *Dirt Roads to Dixie*, 164–65. Flink, *America Adopts the Automobile*, 109–12.

61. Hugh Chalmers, "Relation of the Automobile Industry to the Good Roads Movement," 142–49.

62. Seely, *Building the American Highway System*, 24.

63. Ibid., 44.

64. Congressman W. P. Brownlow was from Tennessee, and Senator Asbury Latimer was from South Carolina. Both proposed similar bills a couple of years earlier, which were later merged into the Brownlow-Latimer bill. The Brownlow bill was H.R. 15369, 57th Congress (1902), and the Latimer bill was S. 3477, 58th Congress (1904). On support for the bill, see U.S. Senate Committee on Agriculture and Forestry, *Roads and Road Building Hearing*; "Good Roads in Sight," *Motor Age* 5, no. 8 (February 25, 1904): 20; and "A Common Duty," *Southern Planter* 66 (July 1905): 564.

65. Senator Candler, "Road Improvements," *Congressional Record* 48, pt. 1 (December 19, 1911): 525–27.

66. H.R. 11686, 63rd Cong. (1914). For debates and the roll-call vote, see "Rural Post Roads," *Congressional Record* (February 10, 1914): 3272–92. The bill did not find a sponsor in the Senate. See Seely, *Building the American Highway System*, 41.

67. Seely, *Building the American Highway System*, 38–40.

68. Preston, *Dirt Roads to Dixie*, 30. Two hundred seventy-four, or 60 percent, of the nation's good-roads organizations were in the South.

69. Hilles, "The Good Roads Movement," 58–62; Seely, *Building the American Highway System*, 27.

70. Clifford Anderson to M. O. Eldridge, June 19, 1902, Box 49, Folder "135 Ga. Roads, 1902–1912," BPR.

71. Clifford L. Anderson to Logan Waller Page, March 12, 1906, Box 49, Folder "135 Ga. Roads, 1902–1912," BPR.

72. J. F. Surrency to Secretary of Agriculture, July 7, 1906, Box 49, Folder "135 Ga. Roads, 1902–1912," BPR.

73. Sydney Clare to Secretary of Agriculture, July 9, 1906, Box 49, Folder "135 Ga. Roads, 1902–1912," BPR.

74. E. L. Bardwell to Secretary of Agriculture, July 13, 1909, Box 49, Folder "135 Ga. Roads, 1902–1912," BPR.

75. R. P. Hall to OPR, December 8, 1908; Director to R. P. Hall, December 12, 1908; Director to R. P. Hall, May 17, 1909; all in Box 49, Folder "135 Ga. Roads, 1902–1912," BPR.

76. Seely, *Building the American Highway System*, 27.

77. A. S. Cushman to Congressman Thomas M. Bell, January 30, 1909, Box 49, Folder "135 Ga. Roads, 1902–1912," BPR. For other examples, see Robert J. Downey to Honorable Charles G. Edwards, November 6, 1911, Box 49, Folder "135 Ga. Roads, 1902–1912," BPR; and Max L. McRae to Honorable W. G. Brantley, March 6, 1911, Box 49, Folder "135 Ga. Roads, 1902–1912," BPR.

78. Joel T. Deese to Dudley Hughes, October 24, 1909; Dudley M. Hughes to Joel T. Deese, November 16, 1909, Congressional File, Political Series, Box 11, Folder 13, Dudley Mays Hughes Collection, Richard B. Russell Library for Political Research and Studies, University of Georgia.

79. Flink, *America Adopts the Automobile*, 322–28. Goddard, *Getting There*, 57–58.

80. Goddard, *Getting There*, 57–58. For a more detailed history of the ARM, see "Meeting of American Road Builders Association," *Farm Equipment Dealer* 9 (October 1911): 86.

81. Seely, *Building the American Highway System*, 33–39.

82. Flink, *America Adopts the Automobile*, 36–47; Flink, *The Automobile Age*, 32–33; and Flink, *The Car Culture*, 22–24.

83. Preston, *Dirt Roads to Dixie*, 48–49. "Glidden Tourists Start South To-Day," *NYT*, October 14, 1911, 14.

84. "Glidden Tour Cars Ditched in Virginia," *NYT*, October 18, 1911, 9. Cox, *Dreaming of Dixie*, 140–42.

85. "Glidden Tourists Swamped in Creek," *NYT*, October 19, 1911, 10.

86. "Governors Discuss Building Highways," *NYT*, October 29, 1911, C8.

87. "Auto Tourists in Peril," *NYT*, October 28, 1909; "Autoists at Roanoke," *NYT*, October 29, 1909.

88. "The Great National Highway," *Southern Good Roads*, January 1910, 15.

89. Preston, *Dirt Roads to Dixie*, 42–48.

90. Cox, *Dreaming of Dixie*, 143.

91. Grantham, *Hoke Smith and the Politics of the New South*.

92. Address by Hoke Smith, *Proceedings of the Fourth American Road Congress*, 17–20.

93. Address by George Diehl, *Proceedings of the Fourth American Road Congress*, 24–27.

Chapter 2

1. "Hoosier Motor Clubs Want South'n Highway," *RTH*, November 29, 1914, 1. "Great Highway Is Planned from Chicago to Florida," *AC*, November 10, 1914, 1. "A Highway to the West," *AC*, November 11, 1914, 8. The name "Dixie Highway" was not used regularly until early 1915.

2. Ralston quoted in "Governors' Meeting Called to Make Plans for Great Highway from Chicago to Dixie," *AC*, December 13, 1914, 4A. "The Dixie Peaceway," *NYT*, April 4, 1915, C2.

3. Preston, *Dirt Roads to Dixie*, 61, 129–30.

4. "The Great National Highway," *Southern Good Roads*, January 1910, 14–16. The Broadway-Whitehall Highway was also known as the National Highway, not to be confused with the nineteenth-century National Road or Cumberland Road.

5. "A Great Highway Connecting the South with the North and West Is Needed," *Manufacturers Record*, September 25, 1913, 45–46. The *Record* is also quoted in Preston, *Dirt Roads to Dixie*, 50.

6. Fisher was among many good-roads boosters who argued that the Lincoln Highway was a more-fitting memorial to the fallen president than a monument in Washington, D.C., for which Congress approved plans in 1913. Their entreaties failed to generate congressional support, despite being read into the *Congressional Record* by a sympathetic ally. See Extension of Remarks of Frank Buchanan of Illinois in the House of Representatives, *Congressional Record* 48, pt. 12 (August 17, 1912): 597–606.

7. While the LHA did not plan to build the highway entirely with private capital, they hoped it would facilitate the process. Hokanson, *The Lincoln Highway*, 5–20. Foster, *Castles in the Sand*, 100–13. "Goodyear Contributes to Road Fund," *Motor Age* (October 3, 1912): 14.

8. Hokanson, *The Lincoln Highway*, 4, 17–20. Post was a well-known journalist and magazine essayist who would become more famous in the 1920s and 1930s as an etiquette columnist. Although the LHA continued to meet for years thereafter, in the immediate aftermath of the routing controversy, the organization withdrew from major publicity stunts like the 1913 tour; refused to grant any more requests for extensions;

stopped soliciting huge private donations; and allowed local, state, and, ultimately, federal government to finish what it had started. Portions of the Lincoln Highway, like the Dixie Highway, were later incorporated into state and federal highway systems. See Hokanson, *The Lincoln Highway*, 15–30, 109.

9. "A Great Highway Connecting the South with the North and West Is Needed." Preston, *Dirt Roads to Dixie*, 51–52.

10. Jane Fisher, *Fabulous Hoosier*, 21, 28–29, 93–106.

11. Over the years, Fisher marketed and sold homes to many of the wealthiest and most prominent members of the automobile industry. See Box 1, Folders 11–12, Carl G. Fisher Collection, Historical Museum of South Florida, Miami (now HistoryMiami).

12. Foster, *Castles in the Sand*, 119–21. Preston, *Dirt Roads to Dixie*, 50–53.

13. "Great Highway Is Planned from Chicago to Florida," 1. Previous meetings were held in Richmond, Va.; Atlantic City, N.J.; and Detroit, Mich. Samuel M. Ralston to John M. Slaton, Nov. 6, 1914, DOC 3085, Record Group 1, Series Group 1, Series 5, Executive Department Correspondence: John Marshall Slaton, Box 218, GDAH. All folders of Slaton's correspondence are labeled by dates of letters rather than numbers.

14. "Hoosier Motor Clubs Want South'n Highway." "The Chicago-Miami Highway," *Miami News*, December 11, 1914, 6. "Great Highway Is Planned from Chicago to Florida." Even as William Gilbreath was making the Dixie Highway announcement in Atlanta, Fisher and the Lincoln Highway were in the headlines back home in Indianapolis, although Fisher had little to do with the LHA by this time. See "Carl G. Fisher Is Elected Lincoln Highway Official," *Indianapolis Star*, November 11, 1914, 9. Fisher's hometown paper did not cover Gilbreath's announcement.

15. As noted previously, the name "Dixie Highway" was not used regularly until early 1915. William Gilbreath to John Marshall Slaton, December 14, 1914; Clark Howell to John Marshall Slaton, December 17, 1914; both in DOC 3086, RG1/SG1/S5, Box 219, Executive Department Correspondence: John M. Slaton, GDAH. "Governors' Meeting Called to Make Plans for Great Highway from Chicago to Dixie."

16. "Work for the Dixie Highway," *LE*, April 26, 1915, 2.

17. *Proceedings of the Fourth American Road Congress*, 24–45. For an example of an appeal for state aid, see A. N. Johnson, "State Control of Road Construction," *Proceedings of the Fourth American Road Congress*, 117–22.

18. "Great Highway Is Planned from Chicago to Florida." "Congress to Ask Government Aid for Good Roads," *AC*, November 10, 1914, 1. "Governors' Meeting Called to Make Plans for Great Highway from Chicago to Dixie."

19. W. W. Fuller to Governor John M. Slaton, December 17, 1914, Executive Department Correspondence: John Marshall Slaton, DOC 3086, RG1/SG1/S5, Box 219, GDAH.

20. W. W. Davis to Governor Tom C. Rye, December 25, 1914, GP 37, Microfilm Roll 12, Box 26, Folder 9, TRP.

21. Charles E. James to John M. Slaton, December 17, 1914, DOC 3086, RG 1, Series 5, Executive Department Correspondence: John M. Slaton, Box 219, GDAH.

22. A. A. Womack (on behalf of Manchester Commercial Club) to Governor Tom C. Rye, March 9, 1915, GP 37, Microfilm Roll 12, Box 26, Folder 9, TRP. Davis and Womack were advocating the same route between Nashville and Chattanooga.

23. Remarks by Stanton Warburton, 62nd Cong., 3rd sess., *Congressional Record*, Appendix, 150–57. Seely, *Building the American Highway System*, 40–41.

24. E. W. James to Vernon M. Peirce, March 10, 1912, Box 49, Folder "135 Ga. Roads, 1902–1912," BPR.

25. Ohio also had a state highway commission, but at the time of the proposal, it was not clear whether the state would be on the route. On the status of state highway commissions, see "Directory of State Highway Officials," *Good Roads* 46 (April 4, 1914): 214–16; and "Highway Laws of the United States," *Good Roads* 46 (April 4, 1914): 217–30.

26. Both refused to commit to attending and cited the possibility of conflicting engagements. John M. Slaton to William Gilbreath, December 16, 1914, and Slaton to Clark Howell, December 18, 1914; both in DOC 3086, RG 1 Series 5, Executive Department Correspondence: John M. Slaton, Box 219, GDAH. Thomas C. Rye to A. A. Womack, March 13, 1915, GP 37, Microfilm Roll 12, Box 26, Folder 9, TRP.

27. Remarks by Governor John M. Slaton, *Proceedings of the Fourth Annual American Road Congress*, 8–9.

28. Remarks by Mrs. John M. Slaton, *Proceedings of the Fourth American Road Congress*, 337–38.

29. Remarks by Senator Hoke Smith, *Proceedings of the Fourth American Road Congress*, 17–20.

30. Remarks by Clarence Kenyon, *Proceedings of the Fourth American Road Congress*, 41–45.

31. Road statistics cited in speech by Commissioner James H. MacDonald of Connecticut, *Proceedings of the Fourth Annual American Road Congress*, 360; Remarks by Governor John M. Slaton, *Proceedings of the Fourth Annual American Road Congress*, 8–9. Slaton subsequently pardoned Leo Frank, which led to Frank's lynching at the hands of a well-organized white mob. See Oney, *And the Dead Shall Rise*.

32. Sanders, *Roots of Reform*, 226–27, 333–35. For a thorough study of the role of home-demonstration agents in the rural farm economy, see Jones, *Mama Learned Us to Work*.

33. Link, *The Paradox of Southern Progressivism*, 262–73, 283–95, 304–11.

34. "Great Highway Is Planned from Chicago to Florida."

35. *Illinois Highways* 2, no. 1 (January 1915): 2–16; *Illinois Highways* 2, no. 2 (February 1915): 18–19; *Illinois Highways* 2, no. 3 (March 1915): 34–37, 41–44. On the background of the Illinois highway department, see Hugill, "Good Roads and the Automobile," 329–30; and Wrone, "Illinois Pulls out of the Mud," 54–61. Illinois was praised at the time for having one of the strongest highway commissions in the nation. See A. N. Johnson, "State Control of Road Construction," *Proceedings of the Fourth Annual American Road Congress*, 121–22.

36. Suzanne Fischer, "The Best Road South: The Failure of the Dixie Highway in Indiana," in *Looking beyond the Dixie Highway*, ed. Stager and Carver, 1–15. Hugill, "Good Roads and the Automobile," 338–39. Starting in 1913, Indiana collected hundreds of thousands of dollars in auto licensing fees, but it is unclear whether any of this money went to road construction or maintenance. See Phillips, *Indiana in Transition*, 265–66.

37. Wright, *The Dixie Highway in Illinois*, 7–9. "Will Plan Today Great Highway," *NYT*, April 3, 1915, 7.

38. Johnson, "State Control of Road Construction," *Proceedings of the Fourth Annual American Road Congress*, 118. Hugill, "Good Roads and the Automobile," 338–39. Kentucky led the South in mileage of improved roads. See "A Lot of Room for Improvement," *LE*, April 7, 1915, 2.

39. "Commercial Organizations and Good Roads," *Worcester Magazine*, November 1915, 260–61. "Will Plan Today Great Highway."

40. "Great Highway Is Planned from Chicago to Florida." "Governors' Meeting Called to Make Plans for Great Highway from Chicago to Dixie."

41. "Commissioners Meet in Chattanooga Today." Hugill, "Good Roads and the Automobile," 339. Martha Carver, "Drivin' the Dixie Highway in Tennessee," in *Looking beyond the Dixie Highway*, ed. Stager and Carver, 20–24. Eastern Tennessee counties passed some of the state's biggest bond issues, in part because of smaller populations and insufficient tax bases, but the money did not go far there; road construction and maintenance were more expensive due to the difficult geography. Of the thirteen counties with a sizable proportion of improved roads, seven were in middle Tennessee, one was in west Tennessee, and six were in east Tennessee. Difficult terrain in east Tennessee mitigated the effect of bond revenues there. See "Road Conditions, Improvements and Bond Issues in the Counties of the State," *Tennessee Agriculture* 4, no. 1 (January 1915): 6–16; and M. O. Eldridge, "The Road Situation in Tennessee," *Tennessee Agriculture* 4, no. 1 (January 1915): 20.

42. J. W. Clark to Governor Tom C. Rye, March 26, 1915, GP 37, Microfilm Roll 12, Box 26, Folder 9, TRP. There were some rivalries within rivalries: among supporters of the Nashville option, there were disagreements over the exact route the highway should take through middle Tennessee. One businessman claimed that the middle Tennessee route passed through "a desert land, for miles at a time without the sign of human beings, a road through a farming country practically worthless" and urged the governor to select a route that detoured through his hometown instead. See A. A. Womack to Governor Tom C. Rye, March 9, 1915, TRP.

43. "Will Plan Today Great Highway," 7. Charlie James, a wealthy Chattanooga businessman who owned land along the third route, offered to subsidize the cost of construction. See "Two Route Are Proposed for Highway in Tennessee," *AC*, March 30, 1915, 7.

44. *Thirteenth Census of the United States Taken in the Year 1910*, vol. 2, *Population*. Florida was the largest state in land area. After Michigan was added to the route during the routing meeting in May 1915, Georgia became the third largest in area.

45. C. M. Strahan, "Why Georgia Builds Top Soil and Sand-Clay Roads," *Proceedings of the Fourth Annual American Road Congress*, 329–30. These figures exclude the seven most urban counties in the state, which had denser populations and more tax revenues. Sheffield, "Improvement of the Road System of Georgia," 22.

46. "Georgia's Rural Route Mileage," *Albany Herald*, November 6, 1914, 1.

47. The section up for grabs did not actually end in Atlanta but in a small town just north of Atlanta.

48. Dalton's population was 5,324 in 1910 and 5,222 in 1920, while Rome's was 12,099 in 1910 and 13,252 in 1920. *Fourteenth Census of the United States*, vol. 1, *Population*, 192–94. On the history of both towns, see Battey, *A History of Rome and Floyd County*, chaps. 1–5; Whitfield-Murray Historical Society, *An Official History of Whitfield County, Georgia, 1852–1999*, 27–30, 67–71; and Flamming, *Creating the Modern South*, 11–22. On the Johnston-Sherman Highway, see Caudill and Ashdown, *Sherman's March in Myth and Memory*, 164–65; and "Whitfield and the Highway," *AC*, January 17, 1915, 4F. The Johnston-Sherman Highway was actually the focus of another routing campaign between Rome and Dalton in 1911 until engineers decided to route the highway through Dalton, closer in line with the actual paths that Johnston and Sherman had taken in 1864. See "Rome May Get on Proposed Highway," *RTH*, September 3, 1911, 1; and "Gordon County Won't Help Build Highway," *RTH*, October 16, 1913.

49. Caudill and Ashdown, *Sherman's March in Myth and Memory*, 164. In 1915 the Dalton route consisted of just three counties—Catoosa, Whitfield, and Gordon—because farther south, both the Battlefield Route and the Rome route merged in Bartow County, approximately forty miles north of Atlanta.

50. U.S. Department of the Interior, *Laws Relating to the National Park Service*, 227–58. "Chattanooga Is Preparing for Dixie Highway Meeting," *AC*, May 13, 1915, 8. "Highway Boosters File Their Data with Commission," *AC*, May 11, 1915, 7. "The Dixie Highway," *Calhoun Times*, May 6, 1915, 4.

51. Caudill and Ashdown, *Sherman's March in Myth and Memory*, 164. "Whitfield and the Highway," *AC*, January 17, 1915, 4F. "Dalton and Whitfield County Set Prosperity Pace," *AC*, January 17, 1915, 7.

52. On Allen's family background, see finding aid to Allen Family Papers, MSS 1014, Atlanta History Center. Later that year, Stone Mountain would serve as the site of the revival of the Ku Klux Klan. Allen actually claimed to have been the first person in Atlanta to speak to William Gilbreath about the Dixie Highway. See Ivan E. Allen to Governor Thomas C. Rye, April 10, 1915, GP 37, Microfilm Roll 12, Box 26, Folder 9, TRP.

53. "Good Roads Clans and Developers from Illinois to Florida Preparing for Epochal Meeting in Chattanooga," *AC*, March 28, 1915, 5F. "Dalton's Claim Is Backed by *Atlanta Constitution*," *North Georgia Citizen*, January 21, 1915, 1. "Information Relative to Route to Atlanta," *North Georgia Citizen*, March 11, 1915, 1.

54. "Committee in Atlanta to Urge Rome Route for Dixie Highway," *AC*, April 28, 1915, 10.

55. "Council Calls Bond Election for Public Improvements," *Rome News-Tribune*, December 22, 1914, 1; and "The Committee, Appointed by Major, Recommends a Bond Issue of $100,000," *Rome News-Tribune*, December 24, 1914, 1.

56. "Dalton and Whitfield County Set Prosperity Pace," *AC*, January 17, 1915, 7. "Gordon County Won't Help Build Highway," *RTH*, October 16, 1913. "Tyler, of Dalton, Good Roads Expert, Talks 'Dixie' Highway," *AC*, March 11, 1915, 6. "Better Roads for Gordon County," *Calhoun Times*, April 8, 1915, 1. Gordon County also tried but failed to change its statute labor law during the routing campaign. See "Road Law Election Postponed Indefinitely," *Calhoun Times*, April 29, 1915, 1.

57. Minutes of the Floyd County Commissioners, March 1, 1915, Book 12, 56–60, RFRC.

58. "Roads in Bad Condition," *AC*, January 3, 1915, 7B. Baker, *Chattooga County*, 63–64, 70–71.

59. "Pike Names Committee to Work for Highway," *AC*, April 23, 1915, 9; "Dekalb, Clayton, Henry, Butts, Monroe Organize to Secure Dixie Highway," *AC*, April 27, 1915, 9; "Counties Striving for Dixie Highway," *AC*, April 30, 1915, 8; "New Route Urged Atlanta to Macon," *AC*, May 1, 1915, 5.

60. "7 Counties Organize to Get Dixie Highway," *AC*, April 16, 1915, 6; "Fitzgerald-Waycross Route Given Boost," *AC*, April 28, 1915, 10; "Savannah Man Urges Old Capital Route," *AC*, May 4, 1915, 6; "Says This Is the Short Route," *AC*, May 6, 1915, 6; "12 Counties Indorse Route Thro' Savannah," *AC*, May 8, 1915, 7.

61. Doyle, *New Men, New Cities, New South*, 15.

62. Between 1880 and 1910, Savannah's population more than doubled, but Macon's more than quadrupled. Between 1910 and 1920, both cities grew at a slower but comparable pace, a testament to the importance of highways like the Dixie. See *Compendium of the Tenth Census of the United States*, pt. 1, 85–86; and *Fourteenth Census of the United States*, vol. 1, *Population*, 193–94.

63. Holmes, *Those Glorious Days*. Durden, *A History of Saint George Parish, Colony of Georgia, Jefferson County, State of Georgia*, 11–34. Knight, *Georgia's Landmarks, Memorials, and Legends*, 157. Georgians migrated farther and farther inland as the state acquired territory from the Creek Indians. See Hudson, *Creek Paths and Federal Roads*, 30–33.

64. Bonner, *Milledgeville*.

65. On the decline of roads in the region in the nineteenth century, see Southerland and Brown, *The Federal Road through Georgia, the Creek Nation, and Alabama, 1806–1836*, 135–43.

66. "Dixie Highway Organized," *AC*, April 4, 1915, 4F. "Will Plan Today Great Highway," 7. "Commissioners Meet in Chattanooga Today." "National Highway May Come through This City," *Union Recorder*, May 4, 1915, 1.

67. These apocryphal stories have become part of the legend of Sherman's March to the Sea. See, for example, Dunkelman, *Marching with Sherman*, 11–12. Melton, "The Town That Sherman Wouldn't Burn," 201–30.

68. "Savannah Man Urges Old Capital Route," *AC*, May 4, 1915, 6.

69. The Black Belt encompassed approximately three dozen small counties stretching horizontally across central Georgia and southwest Georgia.

70. Bartley, *The Creation of Modern Georgia*, 43. Formwalt, "A Garden of Irony and Diversity," 499–503, 508–9. Coleman, *A History of Georgia*, 261.

71. The boll weevil reached Georgia by 1914 but did not threaten the state's cotton crop until 1919. See Giesen, *Boll Weevil Blues*, 142–46.

72. Historians have underestimated the role that transportation and marketing networks played in the South's slow transition to diversified crops. Most have focused on other economic, environmental, and cultural factors, such the prevalence of the

crop-lien system, low yields for food crops, and a resistance to change among conservative and traditionalist rural southerners. See, for example, Aiken, *The Cotton Plantation South since the Civil War*; Daniel, *Breaking the Land*; Fite, *Cotton Fields No More*; Wright, *Old South, New South*; Range, *A Century of Georgia Agriculture*, 96–109; and Giesen, *Boll Weevil Blues*, 109. Range briefly mentions the difficulty of marketing perishable food crops (page 99). Giesen acknowledges the importance of infrastructure to Mississippi cotton farmers but does not elaborate on it (page 60). The role of infrastructure in limiting diversification options is the focus of an unpublished paper: Tammy Ingram, "Distribution before Diversification: Good Farms, Bad Roads, and Agricultural Reform in South Carolina in the 1920s."

73. Sullivan, *Georgia: A State History*, 73.

74. Hughes quoted in "What Georgia Must Do: Diversify Business As Well As Farming," *Atlanta Journal*, September 27, 1914. Clipping in Series IV, Subseries B, Box 6, Folder 8, Dudley Mays Hughes Collection, Richard B. Russell Library for Political Research and Studies, University of Georgia. Ayers, *The Promise of the New South*, 191.

75. Hahamovitch, *The Fruits of Their Labor*, 23–25. Hahamovitch notes that the primary obstacle to widespread truck farming in the South was the lack of capital.

76. Range, *A Century of Georgia Agriculture*, 110–17, 191–94. Georgia Department of Agriculture, *Georgia: The Empire State of the South*, 27–32 (see page 12 for a longer list of other new crops, including sweet potatoes, grain, and corn). On the persistence of cotton farming, see Giesen, *Boll Weevil Blues*, 143–44.

77. "Macon to Thomasville Route Will Have Many Supporters," *AC*, April 3, 1915, 7. "Railroads, Transportation of Perishable Products," Bill no. 219, *Acts and Resolutions of the General Assembly of the State of Georgia*, 84–85. Adams, "Marketing Perishable Farm Products." Range, *A Century of Georgia Agriculture*, 110–11.

78. Williford, *Americus through the Years*, 250. Sawmills and turpentine distilleries grew in southwest Georgia after the Civil War thanks to railroads. See Range, *A Century of Georgia Agriculture*, 155–56. Americus was also a market for grain, oats, wheat, corn, and other diversified crops. See "Striking Evidence of Diversified Farming in Sumter County," *AC*, May 7, 1915, 8.

79. Georgia Department of Agriculture, *Georgia: The Empire State of the South*, 28–32. Range, *A Century of Georgia Agriculture*, 193–94, 216.

80. "Strong Delegation of Dixie Highway Boosters Will Attend Meeting in Americus," *Albany Herald*, April 14, 1915, 2. "Interest in the Dixie Highway Growing Tense," *Albany Herald*, April 12, 1915, 1.

81. "The Dixie Highway," *AC*, April 15, 1915, 8.

82. Fussell, "Touring Southwest Georgia," 552–54. William Rogers, *Thomas County, 1865–1900*, 131–54.

83. "Americus Wants Route," *AC*, March 30, 1915, 7.

84. "South Georgia Sections Vying with Each Other for Great Dixie Highway," *AC*, April 11, 1915, 10B. "Interest Growing in Dixie Highway," *AC*, April 10, 1915, 14. The population of Columbus was 20,554 in 1910 and 31,125 in 1920. *Fourteenth Census of the United States*, 192.

85. "Boosters for Highway Hold Meeting at Camilla," *AC*, April 14, 1915, 4. "Strong

Delegation of Dixie Highway Boosters Will Attend Meet in Americus," 2. "More Than 600 People Dixie Highway Boosters [*sic*] Attended Meet in Americus Today," *Albany Herald*, April 15, 1915, 1.

86. "Meets at Albany and Valdosta Indorse Routes for Highway," *AC*, April 23, 1915, 5.

87. Ray, *Ecology of a Cracker Childhood*, 99–101. Wetherington, *The New South Comes to Wiregrass Georgia*, 47–66, 76–138. Wetherington notes that after 1920, the population of the wiregrass region of southeast Georgia declined dramatically (307–8).

88. "Chamber of Commerce Gets behind the New Highway," *Waycross Journal-Herald*, February 13, 1915, 1; "Wiregrass Counties Form Organization for Highway," *AC*, April 2, 1915, 11.

89. "Burying Money on Roads," *Waycross Journal-Herald*, March 22, 1915, 2. On roads in the wiregrass part of southeast Georgia, see Wetherington, *The New South Comes to Wiregrass Georgia*, chap. 3.

90. "Wiregrass Counties Form Organization for Highway." "Fitzgerald-Waycross Route Given Boost."

91. "Wiregrass Counties Form Organization for Highway." "Five Hundred Highway Delegates Are Present at Fitzgerald Meeting," *AC*, May 1, 1915, 5. "Dixie Highway Meeting Called for Next Friday," *LE*, April 26, 1915, 1.

92. "Dixie Highway," *LE*, April 12, 1915, 1.

93. "Good Weather Aids in Road Work in County," *Waycross Journal-Herald*, March 27, 1915, 1.

94. "Wilcox County Preparing for Dixie-Hoosier Highway," *LE*, April 19, 1915, 1.

95. Nelson, *Trembling Earth*, 104.

96. "Much Damage Is Done by Rain," *Waycross Journal-Herald*, January 19, 1915, 1.

97. "Wilcox County Preparing for Dixie-Hoosier Highway."

98. "Good Weather Aids in Road Work in County." "Ware County Sure to Work for Highway," *Waycross Journal-Herald*, April 5, 1915, 1.

99. "Fitzgerald and Ocilla's Duty," *LE*, April 9, 1915, 4.

100. "More Than a Single Road," *AC*, May 13, 1915, 8.

101. "South Georgia Sections Vying with Each Other for Great Dixie Highway." "An Epoch-Making Project," *AC*, May 10, 1915, 6.

102. "Minutes of the Conference of Governors of the States of Illinois, Indiana, Ohio, Kentucky, Tennessee, Georgia and Florida, or Their Representatives Held at the City Auditorium Chattanooga, Tennessee," April 3, 1915, DHAMB, 1. "Will Plan Today Great Highway."

103. "Auto Men Enthuse over Highway," *AC*, March 23, 1915, 8; "Good Roads Clans and Developers from Illinois to Florida Preparing for Epochal Meeting in Chattanooga," *AC*, March 28, 1915, 5F; "Commission Selected by Governors to Pick Dixie Highway's Route," *AC*, April 4, 1915, 1. Fred Houser to Governor John M. Slaton, March 26, 1915, DOC 3089, RG1/SG1/S5, Executive Department Correspondence: John M. Slaton, Box 222, GDAH. Some counties even paid the expenses for their delegates. See Minutes, Meeting of the Floyd County Board of Roads and Revenues, April 1, 1915, RFRC, Book 12, 80.

104. Governor John. M. Slaton to W. R. Bowen, March 19, 1915; Telegram from Slaton to A. A. Lawrence, March 30, 1915, DOC 3089, RG1/SG1/S5, Executive Department

Correspondence: John M. Slaton, Box 222, GDAH. Governor Rye of Tennessee was unsure if he would even attend the meeting. See Rye to A. A. Womack, March 13, 1915, GP 37, Microfilm Roll 12, Box 26, Folder 9, TRP.

105. Charter, Box 4, DHAMB, n.p.; DHAMB, April 3, 1915, 1.

106. "Commissioners Selected by Governors to Pick Dixie Highway's Route," *AC*, April 4, 1915, 5; "New Dixie Highway in State Control," *NYT*, April 4, 1915, 10. "Governors in Full Control," *Chattanooga Times*, April 4, 1915, 1. DHAMB, April 3, 1915, 1.

107. "Times Wasted in Quarrels," *Chattanooga Times*, April 4, 1915, 4.

108. "Governors in Full Control." "Commissioners Selected by Governors to Pick Dixie Highway's Route." "New Dixie Highway in State Control."

109. Guerry, *Men and Vision*, 9–10. The infamous Louisiana demagogue Huey P. Long was one of the company's most famous salesmen. See Savitt, *Disease and Distinctiveness in the American South*, 181; and Hair, *The Kingfish and His Realm*, 51–52.

110. Chandler and Salsbury, *Pierre S. Du Pont and the Making of the Modern Corporation*, 93–120, 260–91, 433–74. Francis and Hahn, *The DuPont Highway*, 7–8. Leynes and Cullison, *Biscayne National Park: Historic Resource Study*, 23–24.

111. "Tyler, of Dalton, Good Roads Expert, Talks 'Dixie' Highway." Preston, *Dirt Roads to Dixie*, 55.

112. "Commissioners Selected by Governors to Pick Dixie Highway's Route." "New Dixie Highway in State Control." DHAMB, April 3, 1915, 1. On rumors about the board of directors, see, for example, W. C. Martin to Honorable John M. Slaton, March 24, 1915; Frank Manly to Honorable John M. Slaton, March 25, 1915; T. S. Shope to Honorable John M. Slaton, Mar. 25, 1915. Slaton acknowledged the rumors as well. See Slaton to Honorable T. S. Shope, Mar. 26, 1915. All in DOC 3089, RG1/SG1/S5, Executive Department Correspondence: John M. Slaton, Box 222, GDAH.

113. DHAMB, April 23, 1915, 3. Collection description for William Arnold Hanger Collection, University of Kentucky Libraries, Special Collections. A couple of years earlier, Hanger had admitted to paying $30,000 to the Tammany Hall political machine in New York for "expert advice" on how to obtain lucrative contracts on the Catskill Aqueduct, but he was never indicted for any crimes. See "Asked Gaffney to Soothe Unions," *NYT*, September 6, 1913, 6. Suzanne Fischer, "The Best Road South: The Failure of the Dixie Highway in Indiana," in *Looking beyond the Dixie Highway*, ed. Stager and Carver, 8–9. F. T. Hardwick to Governor Thomas C. Rye, April 5, 1915; South Chattanooga Business League to Governor Thomas C. Rye, April 7, 1915; Governor Rye to E. G. Tallett et al., April 15, 1915; all in GP 37, Microfilm Roll 12, Box 26, Folder 9, TRP.

114. DHAMB, April 3, 1915, 2.

115. "Dixie's Dynamo Gets Busy for the Big Dixie Highway," *Chattanooga Times*, April 8, 1915, 1. Desmond, *Chattanooga*, 56. On James's real-estate holdings, see www.chattanoogan.com, September 8, 2010.

116. M. M. Allison to Honorable Samuel L. Raulston [*sic*] et al., April 8, 1915; B. W. Landstreet et al. to Honorable Thomas C. Rye, April 8, 1915; C. Buford Payne to Governor Samuel Ralston, April 8, 1915; all in GP 37, Microfilm Roll 12, Box 26, Folder 9, TRP. This folder contains numerous letters of complaint about James's actions.

117. John M. Slaton to Honorable Clark Howell, April 10, 1915, DOC 3090, RG1/SG1/S5, Executive Department Correspondence: John M. Slaton, Box 223, GDAH.

118. DHAMB, April 23, 1915, 3–4.

119. "Battle of Chattanooga," *AC*, April 27, 1915, 8.

120. "Thousand Romans Ready for March on Chattanooga," *AC*, May 17, 1915, 8. "Dalton Has Holiday That Citizens May Go to Highway Meet," *AC*, May 18, 1915, 5. The paper said the Rome caravan was larger, but Dalton supporters returned home with a trophy for having the most automobiles in their caravan.

121. "Dixie Highway Selected in Form of Great Circuit Linking Lakes with Gulf," *AC*, May 23, 1915, 4. DHAMB, May 20–22, 1915; DHAMB, 5–12. At the routing meeting, the directors also voted to incorporate a permanent Dixie Highway Association head-quartered in Chattanooga and elected regular officers to manage it. See By-Laws, DHAMB, 13–20.

122. DHAMB, May 22, 1915, 13–20.

123. DHAMB, June 14th, 1915, 21–22; and "President James Quits Dixie Route," *AC*, June 2, 1915, 1.

124. "James Subscribes $1,000 to Highway," *AC*, June 3, 1915, 9. To smooth things over, James donated $1,000 to the DHA and pledged to remain a founder.

125. DHAMB, June 14, 1915, 21. James also resigned from the board of directors.

126. DHAMB, May 22, 1915, 12.

127. Charter, DHAMB, April 3, 1915, 1.

128. "Work for the Dixie Highway."

129. "Dixie Highway Launched 'mid Wild Enthusiasm," *LE*, April 5, 1915, 1. They planned to deliver the resolution to Congress, but there is no evidence that they ever did.

130. "Millions of Dollars for Dixie Highway," *AC*, July 8, 1915, 6. This article states that the Dixie traversed 499 miles in twenty-four counties in Georgia, but it does not account for the two southeastern branches that had not yet been designated. These branches would safely put Georgia past Michigan's estimated twenty-eight counties on the route and would more than double Georgia's 499 miles of the Dixie Highway.

131. DHAMB, October 11, 1915, 26.

132. DHAMB, March 25, 1916, 31–32.

133. DHAMB, July 1, 1916, 36–37. With the McDonough route off the map, the DHA rerouted the Old Capital Route directly from Atlanta to Savannah, bypassing Macon on that portion of the highway.

134. "Favors Federal Aid for Highways," *AC*, May 26, 1915, 9.

135. "Thousands Attend Highway Meeting," *AC*, July 6, 1915, 9.

136. DHAMB, October 11, 1915, 28.

137. Seely, *Building the American Highway System*, 38–42. This is discussed briefly in chapter 1 as well.

138. H.R. 11686, 63rd Congress (1914). For debates and the roll-call vote, see "Rural Post Roads," *Congressional Record* 51, pt. 4 (February 10, 1914): 3272–92. Seely, *Building the American Highway System*, 39–43. This is also discussed in chapter 1.

139. Seely, *Building the American Highway System*, 41–43. Karnes, *Asphalt and*

Politics, 14–15. On Bankhead's background as a good-roads booster, see Preston, *Dirt Roads to Dixie*, 33–36, 80–88.

140. H.R. 7617, 64th Congress (1916). For debates and the roll-call vote, see *Congressional Record* 53, pt. 2 (January 25, 1916): 1516–37.

141. *Congressional Record* 53, pt. 7 (April–May 1916): 6425–6899, 7119–27, 7459–7571. Bankhead's committee made several amendments to the bill, mostly dealing with strengthening federal oversight of appropriations given to the states.

142. *Congressional Record* 53, pt. 7 (May 8, 1916): 7571. See also Richard F. Weingroff, "Federal Aid Road Act of 1916: Building the Foundation," *Public Roads* 60, no. 1 (Summer 1996). U.S. Department of Transportation, Federal Highway Administration, *America's Highways*, 86. U.S. Senate, Conference Report on Good Roads Bill, to accompany H.R. 7617 (S.doc.474) ,June 26, 1916.

143. "Federal Aid in Construction of Rural Post Roads" (P.L. 64–156) *United States Statutes at Large*, 39 Stat. 355. The bill is also outlined in Seely, *Building the American Highway System*, 41–43; and U.S. Department of Transportation, Federal Highway Administration, *America's Highways*, 86–87.

Chapter 3

1. "Georgia from a Military Standpoint," *DH*, October 1917, 1–2; "Truck Test Results," *DH*, November 1917, 2–5. The test included both the Rome and Dalton routes of the Dixie Highway. Eisenhower served in Georgia for just three months, from September to December 1917. He briefly described his duties at Fort Oglethorpe in his memoir, *At Ease: Stories I Tell to Friends*, 131–32.

2. "History," *DH*, November 1917, 8.

3. Rose, *Interstate*, 8.

4. The stipulations of the Federal Aid Road Act of 1916 were very precise. They could build and improve roads using federal funds, but maintenance was the responsibility of the states.

5. Davies, *American Road*, 5–6, 40–41. Karnes, *Asphalt and Politics*, 44–69.

6. "The Dixie Peaceway," *NYT*, April 4, 1915, C2.

7. Anonymous, "That's What Old Caesar Did," *DH*, November 1917, 16.

8. Stover, *American Railroads*, 167–70. Kennedy, *Over Here*, 252–53.

9. Stover, *American Railroads*, 170–73. Kennedy, *Over Here*, 4–5, 10–11, 252–53.

10. Stover, *American Railroads*, 172–81. The wartime experience induced the railroads to invest in equipment and change operating procedures to ensure greater efficiency and prevent future crises that might require government intervention. Their efforts paid off when World War II again challenged the nation's transportation networks. See Stover, *American Railroads*, 181–91. On the public response to the Railroad Administration, see Kennedy,, *Over Here*, 255.

11. DHAMB, July 1, 1916, 36.

12. Colonel J. P. Fyffe, "Good Roads a War Time Necessity," *DH*, January 1917, 1–2.

13. Ibid.

14. V. D. L. Robinson, "Transportation and the War Game," *DH*, July 1917, 2.

15. Cover page, *DH*, March 1918. "Past, Present, and Future of the Dixie Highway," *DH*, December 1917, 1–7.

16. "Keep Traffic Normal," *DH*, September 1917, 12.

17. Cartoon, *DH*, October 1917, 13.

18. "The Dixie Highway as a Military Road," *DH*, August 1917, 1.

19. "Highways over 100,000 Miles in a Few Years," *Chicago Daily Tribune*, January 28, 1917, D3.

20. "Nation-Wide Liberty Loan Auto Week June 11," *NYT*, June 3, 1917, 83.

21. "Enlarge Federal Road Act," *NYT*, Feb. 24, 1918, 48.

22. "Georgia From a Military Standpoint." Other southern camps were located in cities adjacent to the Dixie Highway, including Charlotte, N.C.; Columbia, S.C.; and Greenville, S.C. A branch of the Eastern Division would later be extended through Greenville.

23. Cover, *DH*, October 1917.

24. "Heavy Motor Use Needs Best Roads," *NYT*, January 27, 1918, 26.

25. "New Cantonments Mean More Roads," *DH*, September 1917, 15.

26. Untitled, *DH*, November 1917, 16.

27. "Lincoln Highway Travel," *NYT*, January 6, 1918, 116. Lincoln Highway supporters called their route "a thorough, connected and well-marked road entirely across the country" but admitted it was "in no sense a boulevard leading from one coast to the other." *Dixie Highway* claimed that only 230 miles of the Dixie had not been built and/or surfaced, but this was a gross exaggeration. Most of the highway was there in some form, as it had been in 1915, but it had not been improved, much less surfaced. Ibid.

28. DHAMB, May 16, 1918, 60–61.

29. DHAMB, May 16, 1918, 60–63.

30. "Business Men Plan War Aid for Nation," *NYT*, September 17, 1917, 17. "American Business for Vigorous War," *NYT*, September 19, 1917, 4.

31. "Business Pledges Its All to Nation," *NYT*, September 22, 1917, 1. "Chamber of Commerce of the United States of America," *Good Roads* 14, no. 12 (September 22, 1917): 165.

32. V. D. L. Robinson, "Waste Is Appalling," *DH*, July 1917, 2–3.

33. Seely, *Building the American Highway System*, 50. Childs, *Trucking and the Public Interest*, 7.

34. "More Industries Testify to the Service of Good Roads and Motor Truck Haulage," *ENR* 79, no. 23 (December 6, 1917): 1048–49.

35. Ibid.

36. Seely, *Building the American Highway System*, 50–51.

37. Cover page, *DH*, July 1918.

38. "Activities of the Chattanooga Auto Club," *DH*, January 1918, 10.

39. Ibid. The news from the State Highway Department may have come close to press time, as it appears in a brief note on the first page of the January 1918 issue, which is titled "Relief in Sight."

40. For examples of "Touring Queries," see the February 1917 and May 1917 issues, which contain letters from Iowa, Texas, Illinois, and Michigan tourists. "Milledgeville,

the Thriving City," "Historic Thomasville," "Jacksonville," "Hill City of Florida," "Fitzgerald, Ga., Is Yankee Town of the South Dixie Highway," *DH*, October 1917, 3–7.

41. "The Motor Tourist and the Dollar Sign," *DH*, February 1917, 4. On Chattanooga Auto Club members' backgrounds, see "Chattanooga Auto Club Personnel," *DH*, January 1918, 6–8.

42. "Activities of the Chattanooga Auto Club," 10.

43. "A. N. Johnson, "Macadam Roads Not Best for War Traffic, Says Engineer," *ENR* 79, no. 16 (October 1917): 755–56. Some experts disagreed. See "Macadam Roads Best for War Traffic," *ENR* 79, no. 11 (September 1917): 481. Hokanson, *Lincoln Highway*, 19. Even on the Lincoln, the paved sections were not very long. Scattered "seedling miles" or demonstration miles constituted most of the hard surfacing on the Lincoln. See Hokanson, *Lincoln Highway*, 157–58. Eisenhower also recalled after a 1919 trip that only the eastern sections of the Lincoln were paved with concrete. See Eisenhower, *At Ease*, 158.

44. "On the Road to Dixie via Concrete," *DH*, April 1917, 4.

45. Universal Portland Cement Company ad, *DH*, April 1917, 22.

46. Dixie Portland Cement Company ad, *DH*, April 1917, 25.

47. R. Stephen Sennott, "Roadside Luxury: Urban Hotels and Modern Streets along the Dixie Highway," in *Looking beyond the Dixie Highway*, ed. Stager and Carver, 115–35. Jeffrey Durbin, "The Dixie Highway in Georgia," in *Looking beyond the Dixie Highway*, ed. Stager and Carver, 45–46. Gas stations and lodging options were still rare in the 1910s, but according to Durbin, the numbers grew substantially over the next couple of decades. Whitfield County (Dalton) alone had twenty-seven gas stations and six tourist cabins by 1940. Cox, *Dreaming of Dixie*, 144–47. Belasco, *Americans on the Road*, 40–69, 74–75. Belasco notes that auto camps became more institutional after 1920, when auto tourism became more popular. See chapters 4 and 5.

48. Hotel Savannah/Hotel Jacksonville ads, *DH*, March 1917, 12; Hotel Patten ad, *DH*, April 1918, 12. The DHA Minutes Book records most meetings at the Hotel Patten.

49. Belasco, *Americans on the Road*, 66, 125–42.

50. Lowry and Parks, *North Georgia's Dixie Highway*, 113. From these humble origins, Dalton later grew into the carpet-manufacturing capital of the world. Flamming, *Creating the Modern South*, 178–80.

51. Shaffer, *See America First*, 3–4, 33–36, 100–101.

52. "Dixie Highway Travel Heavy during Winter," *Chicago Daily Tribune*, February 18, 1917, D9.

53. See examples in Wright, *The Dixie Highway in Illinois*.

54. "Albany—Georgia's Busy Spot," *DH*, January 1917, 14.

55. The DHA promoted Civil War sites along the route, such as Andersonville National Cemetery in Georgia, Sherman's March to the Sea along the Old Capital Route to Savannah, and the Stone Mountain memorial outside of Atlanta. See cover, *DH*, January 1917; and *DH*, March 1917, 2–9. On Civil War tourism in Virginia, see Cox, *Dreaming of Dixie*, 130–35.

56. Karen L. Cox argues that in the early twentieth century, songs about Dixie were very popular among northerners. The songs evoked a popular nostalgia for a region

many of them had never even visited and for an era that had long since passed them by. See Cox, *Dreaming of Dixie*, 10–33.

57. McKenzie, "The Development of Automobile Road Guides in the United States," 1–39. Automobile Blue Books were published for nearly three decades before modern highway maps became more popular in the late 1920s. See, for example, the *Official Automobile Blue Book* of 1901.

58. Ibid., 40.

59. DHAMB, May 21, 1917, 53.

60. Ibid., 29–43.

61. DHAMB, August 25, 1916, 41.

62. DHAMB, May 16, 1918, 58.

63. McKenzie, "The Development of Automobile Road Guides," 42–43.

64. Shaffer, *See America First*, 154–60.

65. William J. Weir, "Road Maintenance in the War Zone of France," *ENR* 79, no. 11 (September 13, 1917): 488–92.

66. "Roads in the War Zone," *DH*, September 1917, 11.

67. Weir, "Road Maintenance," 490.

68. "Good Roads Saved France—What of America?," *DH*, May 1917, 1.

69. Henry B. Joy, "National Highways and Country Roads," *Monographs of Efficiency* 3 (April 1917): 3–4.

70. "Southern Appalachian Good Roads Association," *DH*, November 1917, 7.

71. "The Chamberlain-Dent Highway Bill," *DH*, November 1917, 8; "Substantial Highways a Vital Need," *Cement and Engineering News* 30, no. 10 (October 1918): 19.

72. "The Northville Outer Belt Drive," *DH*, November 1917, 7; "The Chamberlain-Dent Highway Bill," *DH*, November 1917, 8.

73. "Southern Appalachian Good Roads Association," *Good Roads*, October 27, 1917, 226.

74. Advertisement, *ENR* 79, no. 14 (October 4, 1917): 625–26.

75. "Roads and Pavements," *Municipal Journal* 43, no. 21 (November 22, 1917): 518. "Highways Are to Relieve Railroads," *DH*, November 1917, 5.

76. "The First Motor Truck Convoy," *ENR* 80, no. 1 (January 3, 1918): 3. Presumably, the convoy followed parts of the Lincoln Highway, but that is not clear. See also Seely, *Building the American Highway System*, 50. Although the *ENR* does not name Chapin, Seely says that the convoy was his idea. Seely also claims the experiment involved 30,000 trucks, but the article's report of thirty seems more likely.

77. "Motor Trucks to Relieve the Railways Further," *ENR* 80, no. 26 (June 27, 1918): 1252; "The Return Loads Bureau," *DH*, June 1918, 10–11.

78. Untitled, *DH*, January 1918, 27.

79. "Floyd County Forging Ahead," *DH*, April 1917, 18. The article estimates the county road fund to be $100,000, but the county's annual report suggests it was significantly larger. See Minutes of the Floyd County Commissioners, February 6, 1917, Book 12, 377–78, RFRC.

80. Untitled, *DH*, August 1917, 8.

81. "Bond Issue for Thomas County," *DH*, April 1917, 18.

82. DHAMB, May 21, 1917, 48–49.

83. Ibid, 49.

84. "Past, Present, and Future of the Dixie Highway," *DH*, December 1917, 3.

85. Ibid., 2–4.

86. "The Nation's Business [Membership advertisement]," *DH*, March 1918, 13.

87. Inside front covers, *DH*, March–May 1918.

88. Untitled, *DH*, January 1918, 25.

89. Seely, *Building the American Highway System*, 51–53.

90. Ibid., 57. Childs, *Trucking and the Public Interest*, 10.

91. Every state had its own highway department by 1917. See Rose, *Interstate*, 8.

92. The October 1918 and December 1918 covers of *Dixie Highway* featured Florida tourism.

93. Karnes, *Asphalt and Politics*, 42–69. Eisenhower, *At Ease*, 155–66. Davies, *American Road*, 3–6, 40–41, 215–16.

94. Cover page, *DH*, December 1921.

Chapter 4

1. Smith, "Memory of a Large Christmas."

2. Ibid., 104–8.

3. Ibid., 108.

4. Lichtenstein, "Good Roads and Chain Gangs," 369–95.

5. Chain gangs originated around the turn of the century in most states, including northern states, but they lasted longer in the South. Georgia was the last state to outlaw chain gangs in 1945.

6. Oshinsky, *Worse than Slavery*, 7.

7. Ibid., 20–34. For more on prison labor in the South, see Lichtenstein, *Twice the Work of Free Labor*. On nineteenth-century prison reform, see Sullivan, *The Prison Reform Movement*, 1–22.

8. In 1825 Kentucky became the first state to lease prisoners to private individuals, but northern states used the very similar contract system whereby prisoners remained under state control while their labor "was sold to the highest bidder." See Sullivan, *The Prison Reform Movement*, 13.

9. Oshinsky, *Worse than Slavery*, 35–53. While Oshinsky argues that Richardson's 1868 contract marked the inauguration of convict leasing in Mississippi, he says that the system did not "fully take hold" until after 1875, when Democrats regained control of state politics and began a reign of terror against black Mississippians.

10. Lichtenstein, *Twice the Work of Free Labor*, 41–68.

11. Oshinsky, *Worse than Slavery*, 35. For a comparative account of convict labor on railroads in Virginia, see Nelson, *Steel Drivin' Man*.

12. Lichtenstein, *Twice the Work of Free Labor*, 45–72.

13. Oshinsky, *Worse than Slavery*, 76.

14. Sheffield, "Improvement of the Road System of Georgia," 24–29.

15. Holmes is quoted in Lichtenstein, "Good Roads and Chain Gangs in the Progressive South," 87.

16. Roy Stone, "Address at Manassas," 1–7.

17. On Georgia statistics, see Lichtenstein, "Good Roads and Chain Gangs in the Progressive South," 94–99, and Lichtenstein, *Twice the Work of Free Labor*, 175.

18. Lichtenstein, *Twice the Work of Free Labor*, 165–75.

19. Mancini, *One Dies, Get Another*, 222–23. Lichtenstein, *Twice the Work of Free Labor*, 175.

20. "Georgia's Big Task," *Atlanta Georgian*, May 27, 1908, 6. The paper also printed detailed accounts of poor conditions in convict camps. See "Where Men Die Like Dogs," July 13, 1908, 1. Second quote from Lichtenstein, "Good Roads and Chain Gangs in the Progressive South," 89.

21. "Convict Labor and Georgia," *Journal of Labor*, May 15, 1908, 4. Labor advocates opposed any use of convict labor, including chain gangs.

22. Lichtenstein, *Twice the Work of Free Labor*, 150–58. Mancini, *One Dies, Get Another*, 94–98, 223–26.

23. Grantham, *Hoke Smith*, 173–74.

24. Mancini, *One Dies, Get Another*, 215–32. Mancini notes that Alabama was one of the last states to abolish convict leasing, not doing so until 1928, but the same year the state passed new gasoline taxes to fund chain gangs on public roads. Georgia maintained chain gangs until the 1940s.

25. Progressive governors, including Hoke Smith of Georgia, Charles Aycock of North Carolina, and Joseph F. Johnston of Alabama, campaigned on Progressive platforms that included the disfranchisement amendments they shepherded through their state legislatures. Black disfranchisement and the abolition of convict leasing in favor of chain gangs were both authorized in 1908 in Georgia. The linkages between racial control and Progressive reform have been well documented by historians. See Gilmore, *Gender and Jim Crow* (see, in particular, chapter 5). In *Struggle for Mastery*, Michael Perman argues that the reform impulse motivated disfranchisement in only a few southern states, while others, like North Carolina, were responding to the biracial political alliance between populists and Republicans that threatened Democratic hegemony in 1896. Dewey W. Grantham argues that Governor Hoke Smith of Georgia included disfranchisement in his reform agenda for 1908 and saw no inconsistencies between the concept of reform and that of disfranchisement. See Grantham, *Hoke Smith*, 158–59. States such as Ohio and Colorado experimented with using convicts to work the local roads. Ohio authorities, who used black convicts exclusively, were not satisfied with the results. See Marker, "Experimental Road Work in Ohio." Colorado and other western states used convict road labor with more success but were critical of southern states, which, even in 1913, they still associated with convict leases whereby convicts "were sold at auction to the highest bidder." See Tynan, "Prison Labor on Public Roads," 58–60.

26. Lichtenstein, "Good Roads and Chain Gangs in the Progressive South," 105–6.

27. Pratt, "Convict Labor in Highway Construction," 78–87.

28. "Road Building by Convicts Favored by Gov. Slaton," *Chattanooga Times*, April 4, 1915, 5.

29. Quoted in Lichtenstein, "Good Roads and Chain Gangs in the Progressive South," 90.

30. *Second Biennial Report of the Prison Commission*, 1929, 6.

31. Minutes of the Floyd County Board of Commissioners, July 2, 1906, Book 8, 179, RFRC.

32. Steiner and Brown, *The North Carolina Chain Gang*, 6–9.

33. McKelway is quoted in Lichtenstein, "Good Roads and Chain Gangs in the Progressive South," 90.

34. Steiner and Brown, *The North Carolina Chain Gang*, 6.

35. Burns, *I Am a Fugitive from a Georgia Chain Gang!*, 142.

36. Ibid., 48–49, 59.

37. Ibid., 49–53.

38. Ibid., 53–56, 156–57. "Black Betty" was immortalized in a popular folk song written by a Texas prisoner. See Lomax and Lomax, *American Ballads and Folk Songs*, 60–61.

39. Minutes of the Floyd County Board of Commissioners, November 30, 1906, Book 8, 244, and March 1, 1915, Book 12, 54; both in RFRC. The size of the chain gang fluctuated from month to month, but these figures are representative of those years.

40. Joseph S. Turner (chairman of state prison commission) to Governor Smith, September 16, 1908; Smith to the Prison Commission of Georgia, November 21, 1908; both in Series I, Box 22, Folder 4, HSP. For examples of prison commission appeals to punish counties who mistreated prisoners on misdemeanor chain gangs, see Folder 3 in the same series and box.

41. The controversy played out in the local newspaper during the winter of 1917. See "Came to Consider Convict Situation in Floyd County," *RTC*, January 18, 1917, 1; "Chaingang Abolished by County Commissioners," *RTC*, January 28, 1917, 1; R. L. Brown, "Letter to the Editor," *RTC*, January 21, 1917; Scott Davis, "Letter to the Editor," *RTC*, February 4, 1917, 1, 7; Henry Walker, "The County Board," *RTC*, February 23, 1917, 6; "Concerning the Abolished Chaingang," *RTC*, March 6, 1917, 2; "County Chaingang Controversy Subject of Hearing before Judge Bartlett Here Yesterday," *RTC*, March 23, 1917, 1; "Chaingang Retained for Present," *RTC*, March 24, 1917, 1; and J. Scott Davis, "Letter to Editor," *RTC*, March 25, 1917, 1. The papers do not describe the conditions for Penn's reinstatement, but county records reveal that he remained.

42. Minutes of the Floyd County Board of Commissioners, February 6, 1917, Book 12, 377–78, RFRC. Minutes for the years 1908 to 1916 are in books 9–12. Chain gangs actually constituted a larger percentage of the county's expenses prior to 1908, but local tax receipts were considerably smaller. The cost of maintaining the chain gang increased as it grew larger after 1908. See books 7–8 for the years 1903 to 1907.

43. "Employ Free Labor Instead of Convicts," *RTH*, December 21, 1914, 2.

44. "Chairman Scott Davis Tells Cost of Work by Chaingang," *RTH*, February 4, 1917, 1. Davis references the Summerville Road, which the minutes of the board of

commissioners use interchangeably with the Dixie Highway. See, for example, Minutes of the Floyd County Board of Commissioners, June 9, 1915, Book 12, 98, RFRC.

45. Warren Neel, "Report of the State Highway Engineer," *First Annual Report of the Highway Department of Georgia*, 1919, 7–10, GDAH. Neel reported that the state did provide engineering expertise, but this would have been general oversight rather than active involvement in the day-to-day operations of the chain gang. The Highway Commission's observations are on page 57.

46. McCallie, "A Second Report on the Public Roads of Georgia," 10. Lichtenstein, *Twice the Work of Free Labor*, 176–77. Lichtenstein estimates the total number of convicts working the state's roads to be higher, or nearly 4,996 by 1908. Many counties retained statute labor laws after 1908, but there are few references to them in the record. Most likely they were even less vigorously enforced than before. The minutes of the Floyd County Board of Commissioners indicate that the county had a statute labor law in effect until at least 1919.

47. DHAMB, September 1921, 87–88.

48. DHAMB, May 1924, 127–29.

49. DHAMB, May 21, 1917, 48–49; "Past, Present, and Future of the Dixie Highway," *DH*, December 1917, 3. Some of these are discussed in chapter 3.

50. On county involvement in federal aid projects, see *First Annual Report of the Highway Department of Georgia*, 8. For examples of county bonds being used on state and interstate routes, including the Dixie Highway, see DHAMB, May 15, 1919, 68. "Federal Aid for Counties of Georgia in Building Roads Will Be Discussed," *Rome News-Tribune*, April 15, 1917, 1.

51. While larger counties like Floyd County, Georgia, employed engineers, most smaller rural counties did not.

52. "Highway Department Appoints Engineer," *AC*, November 22, 1917, 4; Warren R. Neel, obituary, *Roads and Streets* 84, no. 4 (April 1941): 60.

53. *First Annual Report of the Highway Department of Georgia*, 7–8.

54. *Second Annual Report of the Highway Department of Georgia*, 1920, 81–88, GDAH.

55. Seely, *Building the American Highway System*, 51–59. One proposal in 1919 was called the Townsend Bill, which proposed a federal highway commission. At the time, Congress was embroiled in debates over the Treaty of Versailles, so the bill went nowhere. Congressmen proved unwilling to seriously consider any legislation, in fact, until the 1916 bill expired in 1921.

56. Neel, "Report of the State Highway Engineer," 9. Seely, *Building the American Highway System*, 59. "Layout of Nation-Wide Highway System Needed," *DH*, November 1920, 2.

57. *First Annual Report of the Highway Department of Georgia*, 7–8, 10–20.

58. *Second Annual Report of the Highway Department of Georgia*, 98–105.

59. *Fourth Annual Report of the Highway Department of Georgia*, 1922, 22, GDAH.

60. "What Would a $50,000,000 Bond Issue Do for Tennessee?," *DH*, March 1919, 10–11.

61. "Georgia Unanimous for Bond Issue," *DH*, March 1919, 17; "States Getting in Line," *DH*, June 1920, 1. Minutes of the Floyd County Board of Commissioners, Feb. 5, 1919, Book 13, 31, RFRC.

62. *Second Annual Report of the Highway Department of Georgia*, 21–22.

63. Michael M. Allison to Carl G. Fisher, August 9, 1920, Box 4, Folders 4–6 (microfilm), Carl G. Fisher Collection, Historical Museum of South Florida, Miami.

64. "States Getting in Line." *Third Annual Report of the Highway Department of Georgia*, 1921, 27, GDAH. "Good Roads Bills Passed by Senate at Night Session," *AC*, August 14, 1919, 1. North Carolina was the only southern state to approve bonds. The state passed a $55 million bond in 1921. See *Third Annual Report of the Highway Department of Georgia*, 1921.

65. "Good Roads Bills Passed by Senate at Night Session," *AC*, August 14, 1919, 1. On problems with the older law, see "The Automobile Tax," *AC*, March 28, 1915, 4F. The new bill allowed the state highway department to administer the fund.

66. *Second Annual Report of the Highway Department of Georgia*, 77. The state passed a one-cent gasoline tax in 1921, but revenues were not available for another year.

67. *Third Annual Report of the Highway Department of Georgia*, 24–26.

68. T. E. Patterson quoted in *First Annual Report of the Highway Department of Georgia*, 57. Patterson conceded that counties with small chain gangs could not make them profitable but suggested they combine their road gangs with those of neighboring counties.

69. "Georgia Highway Laws," *Second Annual Report of the Highway Department of Georgia*, 158–68. *Third Annual Report of the Highway Department of Georgia*, 21. On the state's use of convicts, see *Second Annual Report of the Highway Department of Georgia*, 17–18, and Hurst, *This Magic Wilderness*, 458–59.

70. For debates on the bill in the House, see "Construction of Rural Post Roads," *Congressional Record* 61, pt. 3 (June 27, 1921): 3084–94. The House vote was for a slightly modified version of the bill that the Senate had unanimously passed earlier in the year. That bill was authored by Colorado Senator Lawrence C. Phipps. See U.S. Department of Transportation, Federal Highway Administration, *America's Highways*, 108. Phipps worked out his bill as chairman of the Senate Committee on Post Offices and Post Roads (as Bankhead had done in 1916). See U.S. Senate Committee on Post Offices and Post Roads, *Rural Roads* (to accompany S.1072), S.Rp.40 (Washington, D.C.: Government Printing Office, 1921).

71. "An Act to Amend the Act entitled 'An Act to provide that the United States shall aid the States in the construction of rural post roads, and for other purposes,' approved July 11, 1916." (P.L. 67-87), *United States Statutes at Large*, 42 Stat. 212. Overviews of the bill are in Seely, *Building the American Highway System*, 61–65; Rose, *Interstate*, 8–9; and *America's Highways*, 107-8. On Senator Phipps, see Jackson, *The Ku Klux Klan in the City*, 226–28. Phipps was not in the Klan but helped Klan candidates get elected in Colorado.

72. *Third Annual Report of the Highway Department of Georgia*, 20. *Fourth Annual*

Report of the Highway Department of Georgia, 93–94, 141. *Fifth Annual Report of the Highway Department of Georgia,* 3–4, 10–11, GDAH.

73. *Fifth Annual Report of the Highway Department of Georgia,* 73.

74. *Twenty-third Annual Report of the Prison Commission,* 1919, 12–14; *Twenty-fifth Annual Report of the Prison Commission,* 1921, 11–14; and *Twenty-eighth Annual Report of the Prison Commission,* 1925, 11–14.

75. *Third Annual Report of the Highway Department of Georgia,* 35.

76. *Fifth Annual Report of the Highway Department of Georgia,* 9.

77. *Fourth Annual Report of the Highway Department of Georgia,* 44; *Fifth Annual Report of the Highway Department of Georgia,* 2; Federal aid figures in Georgia were very similar to those of the Dixie Highway itself, cited earlier. See DHAMB, May 1924, 127–29.

78. "Nearly 28,000 Miles of Good Roads Built in 1921," *DH,* December 1921, 2.

79. DHAMB, May 1924, 127–34. DHAMB, May 16, 1918, 55.

80. The opposition to state highway bonds will be discussed more fully in chapter 5. For examples of support for county bonds, see "Believed Bond Issue Will Win in Colquitt," *AC,* March 20, 1919, 12; "Baldwin County Wants Bond Issue for Better Roads," *AC,* April 2, 1919, 12; and "Coweta County Votes on $480,000 for Road Bonds," *AC,* April 24, 1919, 15. The state highway department understood this as well and regularly urged citizens to back local bond proposals. See, for example, "Patterson Speaking for County Bonds," *AC,* March 29, 1919, 16.

81. *Sixth Annual Report of the Highway Department of Georgia,* 1924, 11, GDAH.

82. "Tax of One Cent Per Gallon Voted on Gasoline Sale," *AC,* August 6, 1921, 1. "Fourteen States Tax Gasoline," *ENR* 87, no. 12 (1921): 504.

83. "Higher Gasoline Tax Is Proposed," *AC,* March 8, 1923, 7.

84. *Fifth Annual Report of the Highway Department of Georgia,* 1923, 3. The board's proposal was also outlined in the newspapers. See "Plans Are Made for Increasing State's Revenue," *AC,* June 24, 1923, 1.

85. Holder quoted in "Higher Gasoline Tax Is Proposed," *AC,* March 8, 1923, 7.

86. "3-Cent Gasoline Tax Measure Is Passed by House," *Atlanta Journal,* August 7, 1923, 1. "Three Cent Tax Voted for Gas," *AC,* August 8, 1923, 3.

87. "Gas Tax Increase 'Forward Step,' Declares Mann," *AC,* August 19, 1923, 9.

88. *Seventh Annual Report of the Highway Department of Georgia,* 1925, 10, GDAH. *Ninth Annual Report of the Highway Department of Georgia,* 1927, 73, GDAH. The 8th and 9th annual reports were combined into a single volume.

89. "Thirty-Six States Have Already Voted a Gasoline Tax," *Georgia Highways,* July 1924, 22; "Motorists in 36 States Pay $50,000,000 Gas Taxes," *Popular Science Monthly,* January 1925, 71. Among the states without gas taxes was Illinois, which had substantial state highway bonds.

90. *Seventh Annual Report of the Highway Department of Georgia,* 1925, 5.

91. Seely, *Building the American Highway System,* 73. Congress resumed granting the usual $75 million in federal highway aid in 1925.

92. DHAMB, May 1924, 127–29.

93. DHAMB, November 1924, 164–65.

94. *Seventh Annual Report of the Highway Department of Georgia*, 1925, 8.

95. DHAMB, November 1924, 165.

96. "Anderson Named on Highway Board," *AC*, October 30, 1923, 7.

97. *Seventh Annual Report of the Highway Department of Georgia*, 1925, 1.

98. Neel is quoted in Lichtenstein, *Twice the Work of Free Labor*, 190.

99. *Fifth Annual Report of the Highway Department of Georgia*, 1923, 75–77.

100. DHAMB, May 1924, 130–34.

101. *First Biennial Report of the Prison Commission*, 11–14.

102. *Seventh Annual Report of the Highway Department of Georgia*, 1925, 136; *Eighth and Ninth Annual Reports of the Highway Department of Georgia*, 1926–27, IV. These figures did not count Texas among southern states. Texas led Georgia in federal aid mileage. The next chapter will go into greater depth about the state's patchwork roads.

103. *Seventh Annual Report of the Highway Department of Georgia*, 1925, 196–98.

104. "Federal Aid Policy Must Be Continued," *DH*, December 1925, 22.

Chapter 5

1. Fisher, *Fabulous Hoosier*, 79.

2. Anderson, *The Wild Man from Sugar Creek*, 72.

3. Hackbart-Dean, "Georgia's Renaissance Governor." A more-detailed biography of Hardman is located in the finding aid to the LHC.

4. Fite, *Richard B. Russell, Jr., Senator from Georgia*, 65.

5. Preston, *Dirt Roads to Dixie*, 164–66.

6. U.S. Department of Transportation, *America's Highways*, 101–9.

7. Ibid.

8. Daniel, *Breaking the Land*, 18–20.

9. Ibid., 3–22.

10. Monthly Report, "Results of Home Demonstration Work in South Carolina," December 1929, Box 14, Folder 439, Cooperative Extension Service Field Operations.

11. Monthly Report, January 1924, Box 4, Folder 145, Cooperative Extension Service Field Operations.

12. Monthly Report, September 1928, Box 9, Folder 326, Cooperative Extension Service Field Operations; "Result of Home Demonstration Work in South Carolina," December 1928, Box 11, Folder 355, Cooperative Extension Service Field Operations.

13. The 1921 Highway Act would remain in place with few changes until the 1930s. In the intervening years, Congress voted merely to extend the legislation and occasionally increase funding for it rather than alter federal road-building powers. See U.S. Department of Transportation, *America's Highways*, 546–47.

14. "Report of the Committee on Administration of the American Association of State Highway Officials," November 1924, Box 1692, Central File, BPR.

15. Ibid.; Howard M. Gore to Mr. Frank T. Rogers (attached to AASHO "Report"), January 5, 1925, Box 1692, Central File, BPR.

16. Thos. H. MacDonald to Howard M. Gore (attached to AASHO "Report"), February 20, 1925, 5–7, Box 1692, BPR; U.S. Department of Transportation, *America's Highways*, 110.

17. H. J. Carswell to Warren Neel, January 9, 1925, Box 1696, Folder "Georgia," BPR. Executive Committee of the Wilmington Chamber of Commerce to the City and County Governments, South Atlantic Coastal Highway Members and Officials et al., December 21, 1925, Box 1691, Classified Central File, BPR.

18. C. E. Roop et al. to Warren Neel, January 6, 1926, Box 1696, Folder "Georgia," BPR.

19. Neel actually testified in Holder's defense when Governor Clifford Walker accused Holder of misdoings. See "Holder Enters Broad Denial of All Charges," *AC*, August 12, 1925, 1. Holder's controversial reign as state highway chairman will be discussed at greater length in the next section.

20. Warren Neel to W. C. Markham, December 18, 1925; and Markham to Neel, December 21, 1925, Box 1696, Folder "Georgia," BPR.

21. Excerpt from Minutes of meeting of State Highway Board of Georgia, April 14, 1926, Box 1696, Folder "Georgia," BPR. Stanley Bennet, John Holder et al. to E. W. James, June 1926, Box 1696, Folder "Georgia," BPR.

22. Warren Neel to E. W. James, January 23, 1926, Box 1696, Folder "Georgia," BPR. Warren Neel to W. C. Markham, February 24, 1926, Box 1696, Folder "Georgia," BPR.

23. During the campaign, Neel explicitly criticized the "pay-as-you-go plan" advocated by Hardman and, eventually, by his colleague John Holder. See "Highway Engineer Outlines Policy," *AC*, March 19, 1926, 7.

24. Warren Neel to W. C. Markham, February 26, 1926, Box 1696, Folder "Georgia," BPR. State highway routes are clearly marked on highway department maps. See, for example, *Fifth Annual Report of the Highway Department of Georgia*, 1923.

25. Richard F. Weingroff, "From Names to Numbers: The Origins of the Numbered U.S. Highway System," *AASHTO Quarterly*, Spring 1997.

26. DHAMB, October 13, 1925, 187.

27. DHAMB, July 9, 1926, 193.

28. U.S. Department of Transportation, *America's Highways*, 109.

29. *Third Annual Report of the Highway Department of Georgia*, 1921, 50–75. Subsequent annual reports also list state and federal aid projects, but not all of them indicate which ones were also known as the "Dixie Highway."

30. Howard M. Gore to Chairman Dowell, February 9, 1925, Box 1692, Folder "Georgia," BPR.

31. "Resolution No. 5 Regarding Trail Marking," n.d., Box 1696, Folder "Georgia," BPR. Markham refers to this resolution in his letter to Warren Neel on June 19, 1926. Weingroff, "From Names to Numbers."

32. Thomas MacDonald to Honorable William J. Harris, January 11, 1926, Box 1691, Central File, BPR.

33. Weingroff, "From Names to Numbers."

34. Preston, *Dirt Roads to Dixie*, 129–31.

35. DHAMB, September 1, 1926, 194.

36. "Resolution No. 5 Regarding Trail Marking."

37. W. T. Anderson to Mr. T. A. McDonald [*sic*], February 3, 1926, Box 1691, BPR.

38. Letter from Thomas H. MacDonald to Mr. W. T. Anderson, February 10, 1926, Box 1691, BPR.

39. Executive Committee of the Wilmington Chamber of Commerce to the City and County Governments, South Atlantic Coastal Highway Members and Officials et al., December 21, 1925, Box 1691, Classified Central File, BPR.

40. Ibid.

41. S. W. Brown to Honorable C. L. Blease, February 9, 1926, Box 1691, BPR. For an overview of Coleman Blease's early political career, see Simon, *A Fabric of Defeat*, 11–35. In 1924, Blease was elected a U.S. Senator.

42. Fred G. Warde to Senator William J. Harris, January 4, 1926, Box 1691, BPR; William J. Harris to Chief of Bureau of Public Roads, January 6, 1926, Box 1691, BPR. Harris also forwarded protests to changing the U.S. 1 route. See Harris to Chief of Bureau of Public Roads, January 23, Box 1691, BPR; Thomas MacDonald to Honorable William J. Harris, January 11, 1926, Box 1691, BPR. Fred G. Warde to Warren Neel, January 12, 1926, Box 1691, BPR.

43. Fred G. Warde to Warren Neel, January 12, 1926, Box 1691, BPR. Warren Neel to Mr. Fred G. Warde, January 19, 1926, Box 1691, BPR.

44. W. J. Fields to the Honorable Clifford M. Walker, August 13, 1926, Box 1696, BPR; Warren Neel to Mr. W. C. Markham, August 27, 1926, Box 1696, BPR.

45. W. C. Markham to Mr. Warren Neel, August 31, 1926, Box 1696, BPR. Markham responded favorably to Neel's request, but the AASHO did not act on it. U.S. 25 actually overlapped the Carolina Division of the Dixie Highway and connected to the Southeastern Division of the Dixie near Waycross. In Georgia, the eastern portion of the Dixie remained U.S. 41 per the joint board's original decision, while U.S. 25 overlapped a division of the Dixie Highway farther north. Due to changes that occurred later in the twentieth century, however, U.S. 41 currently parallels the approximate original route of the eastern division of the Dixie Highway.

46. Warren Neel to W. C. Markham, May 13, 1926, Box 1696, Folder "Georgia," BPR; W. C. Markham to Warren Neel, June 19, 1926, Box 1696, Folder "Georgia," BPR. On other named routes, see Weingroff, "From Names to Numbers."

47. Warren Neel to W. C. Markham, February 24, 1926, Box 1696, Folder "Georgia," BPR. Neel first suggested marking both routes in his letter to E. W. James of the Bureau of Public Roads on January 23.

48. J. J. Mangham to C. M. Babcock, April 12, 1926, Box 1696, Folder "Georgia," BPR.

49. J. J. Mangham to Honorable Thos. H. McDonald [*sic*], April 29, 1926, Box 1696, Folder "Georgia," BPR.

50. Letter from P. H. J. Wilson to J. J. Mangham, May 5, 1926; Thomas MacDonald to W. C. Markham, April 20, 1926; both in Box 1696, Folder "Georgia," BPR.

51. P. H. J. Wilson to Honorable Gordon Lee, May 4, 1926, Box 1696, Folder "Georgia," BPR.

52. "Gubernatorial Aspirants Stir Simmering Political Pot with Their Statements," *AC*, May 2, 1926; "Wood Launches Race for Governor in Speech to Jasper County

Voters," *AC*, June 10, 1926, 2; "J. O. Wood Challenges Carswell to Debate," *AC*, June 29, 1926, 3.

53. Speech of Dr. L. G. Hardman to the People of Calhoun, Georgia, August 24, 1926, 2, General File, Political Series, LHC. These charges were not new to the public. Holder had been investigated the previous year for improper conduct in office. For an overview of the charges and Holder's response to them, see "Holder Enters Broad Denial of All Charges." Anderson, *Wild Man*, 33–34.

54. Chairmanship of Georgia Highway Commission Offered John N. Holder by Governor Hardwick," *AC*, August 2, 1921, 1.

55. *Eighth Annual Report of the Highway Department of Georgia*, 1926, iii–v, 73, 107, GDAH.

56. "Holder Enters Broad Denial of All Charges." "Conspiracy Charged to Governor Walker in Holder Removal," *AC*, August 13, 1925, 1; "Highway Department Probe Branded by W. T. Anderson as 'Farce and Whitewash,'" *AC*, August 25, 1925, 2.

57. *Seventh Annual Report of the Highway Department of Georgia*, 3. *Eighth Annual Report of the Highway Department of Georgia*, iii–v.

58. "Crazy-Quilt Roads," *AC*, September 27, 1925, D2.

59. "What Are the Unvarnished Facts about Georgia's Highway System?," *AC*, February 1, 1926, 4. "Georgia Roads Being Linked as Rapidly as Funds Allow, Chairman Holder Declares," *AC*, October 4, 1925, A8; "Holder Says State Will Match Bonds," *AC*, December 10, 1925, 8.

60. Charles M. Strahan to Hoke Smith, March 5, 1926, Series I, Box 10, Folder 5, HSP. Newspaper clippings from the *Columbus Enquirer-Sun*, *Thomasville Times Enterprise*, and *Savannah Press*, and *Athens Banner-Herald*, March 1926, Series I, Box 11, Folder 1, HSP. Smith is quoted in the *Banner-Herald* article. Smith received numerous letters from highway progressives urging him to run for governor. See Series I, Boxes 10–11, HSP.

61. "Speaking of Hoke Smith," *Rome News-Tribune*, March 17, 1926, and "Hoke Smith a Parasite," *News Herald* (Lawrenceville, Ga.), March 18, 1926; clippings in Series I, Box 11, Folder 1, HSP.

62. "Wood Continues to Flay Holder," *AC*, July 11, 1926, 11; "Holder Is Flayed by J. O. Wood," *AC*, June 16, 1926, 18; "Carswell Raps Holder's Silence on Road Awards," *AC*, July 29, 1926, 5; "Candidate Carswell Flays John Holder," *AC*, August 4, 1926, 15. Carswell and Wood rarely criticized L. G. Hardman.

63. "Cash Plan for Road Building," *AC*, January 30, 1926, 7.

64. "Address of John N. Holder, Chairman Highway Commission of Georgia, before the Annual Convention of the County Commissioners Association of Georgia, Savannah, Ga.," June 3, 1926, Hargrett Rare Book and Manuscript Library, University of Georgia Libraries, Athens, Georgia.

65. Ibid.

66. Speech of Dr. L. G. Hardman to Cherokee Trail Association, August 23, 1926, General File, Series III, Subseries A, LHC.

67. Speech of Dr. L. G. Hardman to the People of Lexington, Georgia, Series III, Subseries A, LHC.

68. Speech of Dr. L. G. Hardman to the People of Buchanan, Georgia, August 27, 1926, 1, Series III, Subseries A, LHC. "People Are Weary of Detour Signs, Says Hardman," *AC*, August 28, 1926, 5.

69. Speech of Dr. L. G. Hardman to the People of Calhoun, Georgia, Series III, Subseries A, LHC.

70. Speech of Dr. L. G. Hardman to the People of Carrollton, Georgia, August 28, 1926, Series III, Subseries A, LHC.

71. Speech of Dr. L. G. Hardman to the People of Lexington, Georgia, September 20, 1926, Series III, Subseries A, LHC.

72. Statement to *Savannah News, Macon Telegraph, Columbus Enquirer Sun*, October 3, 1926, Series III, Subseries A, LHC.

73. "Hardman Attacks Holder on Highway Bond Issue Record," *AC*, September 26, 1926, 1. The *Athens Banner-Herald*, a newspaper near Holder's hometown of Commerce, also ran a story reminding voters that Holder had once supported the $70 million bond. See "Holder Urged $70,000,000 Bond Issue for Roads," *AC* (quoting Athens article), August 15, 1926, 1.

74. *Sixth Annual Report of the Highway Department of Georgia*, 11; *Seventh Annual Report of the Highway Department of Georgia*, 1.

75. "Holder, in Decatur Speech, Attacks Governor and Heads of Departments of State," *AC*, October 6, 1926, 6.

76. "Wood and Anderson Speak for Hardman," *AC*, October 5, 1926, 8.

77. "George and Talmadge Victors in Primary; Hardman and Holder Lead for Governor," *AC*, September 9, 1926, 1; "The Primary!," *AC*, September 10, 1926, 6.

78. "Dr. L. G. Hardman Wins Race for Governor, Defeating John N. Holder by over 2 to 1," *AC*, October 7, 1926, 1. "Hardman's Total Set at 80,925," *AC*, October 13, 1926, 1. Holder's total was 60,244. Holder was a target of demagogue Eugene Talmadge, a popular and outspoken advocate of Georgia farmers. See "Talmadge Calls on Voters to 'Banish Yoke of Tyrant,'" *AC*, September 8, 1926, 16. Anderson, *Wild Man from Sugar Creek*, 37–49.

79. "Choosing a Governor," *AC*, October 7, 1926, 6.

80. In his place, Hardman nominated the state's original highway chairman, Charles Strahan, despite Strahan's support for state bonds. "Strahan Named State Road Head," *AC*, July 22, 1917, 2.

81. U.S. Department of Transportation, *America's Highways*, 110.

82. "Report of the Committee on Administration of AASHO," 13, and correspondence between Warren Neel and E. W. James, Box 1696, BPR.

83. DHAMB, April 22, 1927, 198.

84. Warren Neel to Mr. E. W. James, February 16, 1927, Box 1696, BPR. The AASHO promised to correct the mistake and add the Rome route to the U.S. highway maps. See E. W. James to Warren Neel, March 5, 1927, Box 1696, BPR.

85. During congressional debates over highway funding in 1925 and 1926, lawmakers cited Coolidge's speeches against federal highway aid. See 68th Cong., 2nd sess., February 3, 1925, *Congressional Record*, 2936; and 69th Cong., 1st sess., March 18, 1926,

Congressional Record, 5886–87. Coolidge is quoted in these debates and in Teaford, *The Rise of the States*, 102.

86. Seely, *Building the American Highway System*, 73–92.

87. Wells, "Fueling the Boom," 75–76. Wells argues that the backlash against gas-tax diversion eventually induced state governments to earmark all gas-tax revenues for highways, which by the 1940s led to a massive "self-replicating" system whereby new highways generated more traffic, and that traffic generated new tax revenues to pay for new highways. This culminated in the creation of the "inviolate" Highway Trust Fund for federal gas-tax revenues in 1956. National Highway Users Conference, *Texts of Good Roads Amendments*.

88. For a comparative chart of federal funding in the 1920s, see *Tenth Annual Report of the Highway Department of Georgia*, 1928, 111, GDAH.

Conclusion

1. Anderson, *The Wild Man from Sugar Creek*, vii, 48–49. "The Primary!," *AC*, September 10, 1926, 6.

2. Anderson, *The Wild Man from Sugar Creek*, 72.

3. Galloway, "Tribune of the Masses," 674–78. Anderson, *The Wild Man from Sugar Creek*, 67–68, 78–79, 82–97.

4. Galloway, "Tribune of the Masses," 678–84; Anderson, *The Wild Man from Sugar Creek*, 83–85. The state legislature refused to pass Talmadge's tag bill, but Talmadge maneuvered around the legislature to impose the reduced fee.

5. Simon, *A Fabric of Defeat*, 167–87.

6. Seely, *Building the American Highway System*, 94.

7. Brinkley, *Voices of Protest*, 72–81.

8. Hamilton, *Trucking Country*.

9. Gutfreund, *Twentieth-Century Sprawl*, 29–32. Rose, *Interstate*, 9.

10. Gutfreund, *Twentieth-Century Sprawl*, 29–30.

11. Preston, *Dirt Roads to Dixie*, 159. Gutfreund, *Twentieth-Century Sprawl*, 30. Seely, *Building the American Highway System*, 84–94.

12. Rose, *Interstate*, 10–12.

13. Rose, *Interstate*, 11–12. Seely, *Building the American Highway System*, 166–77, 192–223.

14. Seely, *Building the American Highway System*, 192–223. Gutfreund, *Twentieth-Century Sprawl*, 37–55. Rose, *Interstate*, 41–54.

15. Seely, *Building the American Highway System*, 192–218; Gutfreund, *Twentieth-Century Sprawl*, 37–55.

16. "Georgia's Rural Roads Program," *Athens Banner-Herald*, August 26, 1958. Clipping in vertical files, Folder "Georgia Roads—General," Hargrett Rare Book and Manuscript Library, University of Georgia Libraries, Athens, Georgia; "Only One-Third of Roads in Georgia Now Paved," *Thomaston Times*, February 4, 1965, vertical files, Folder "Georgia Roads—General," Hargrett Rare Book and Manuscript Library,

University of Georgia Libraries, Athens, Georgia; Phone and e-mail conversations with Bryan Gunter, Georgia Department of Transportation, July 26, 2007.

17. "State Motor Vehicle Registrations, 2005," http://www.fhwa.dot.gov/policy/ohim/hs05/motor_vehicles.htm (retrieved July 26, 2007): "Frequently Asked Questions," http://www.fhwa.dot.gov/policy/ohpi/hss/faqs.htm (retrieved July 26, 2007); "Public Road and Street Length, 1980–2005," http://www.fhwa.dot.gov/ohim/summary95/section5.html (retrieved July 26, 2007).

18. Timothy Egan, "Paying on the Highway to Get out of First Gear," *NYT*, April 28, 2005; Bill Murphy, "Three Studies to Map Future for Toll Roads," *Houston Chronicle*, January 24, 2006.

19. Egan, "Paying on the Highway"; Dave Williams, "State's Transportation Revenue Bypasses Needed Highway Projects," *Athens Banner-Herald*, March 28, 2000.

20. For examples of these debates, see "No Closure for Denver's Beltway Loop," *NYT*, January 16, 2012; "Here We Go Again with Efforts to Extend 526," *Post and Courier* (Charleston, S.C.), June 27, 2012; "Tea Party Notches a Big Win in T-SPLOST Loss," *Atlanta Journal-Constitution*, July 31, 2012; and "Subway Dig OK'd under Beverly Hills High," *Los Angeles Times*, May 25, 2012.

21. "Congress Approves a $127 Billion Transportation and Student Loan Package," *NYT*, June 29, 2012.

Bibliography

Primary Sources

Manuscript and Microfilm Collections

Athens, Ga.
    Richard B. Russell Library for Political Research and Studies, University
        of Georgia
        Dudley Mays Hughes Collection
        Lamartine G. Hardman Collection
        Hoke Smith Papers
    Hargrett Rare Book and Manuscript Library
        John Holder Manuscript
        Vertical Files
Atlanta, Ga.
    Georgia Department of Archives and History
        Annual and Biennial Reports of the Highway Department of Georgia, 1919–1968
        Annual Reports of the Prison Commission of Georgia
        Governor John. M. Slaton Papers
        Road Supervisors Book for Chattooga County
Chapel Hill, N.C.
    North Caroliniana Collection, Wilson Library, University of North Carolina
        North Carolina Good Roads Movement, Newspaper Clipping Scrapbooks
Chattanooga, Tenn.
    Chattanooga–Hamilton County Bicentennial Library
        Records of the Chattanooga Automobile Club & Dixie Highway Association
Clemson, S.C.
    Clemson University Libraries, Special Collections
        Cooperative Extension Service Field Operations, 1909–1985
Dalton, Ga.
    Whitfield-Murray Historical Society
        Photographs

Macon, Ga.
   Washington Memorial Library and Middle Georgia Archives
      Records of the Dixie Highway Auxiliary of Bibb County, 1917–1930
Miami, Fla.
   Historical Museum of South Florida
      Carl G. Fisher Collection
Nashville, Tenn.
   Tennessee State Library and Archives
      Governor Tom C. Rye Papers
Rome, Ga.
   Rome-Floyd Records Center
      Minutes of the Floyd County Commissioners
Washington, D.C.
   National Archives and Records Administration
      Records of the Bureau of Public Roads, Record Group 30

Newspapers and Periodicals

Albany Herald

Athens Banner-Herald

Atlanta Constitution

Atlanta Georgian

Atlanta Journal

Automobile Topics

Automotive Industries

Calhoun Times

Cement and Engineering News

Chattanooga Times

Chicago Daily Tribune

Dalton Citizen

Dixie Highway

Engineering News-Record

Farm Equipment Dealer

Georgia Highways (State Highway
   Department of Georgia, Atlanta)

Good Roads

Houston Chronicle

Illinois Highways (Illinois State
   Highway Department, Springfield)

Indianapolis Star

Journal of Labor

Leader-Enterprise (Fitzgerald, Ga.)

Life

Macon Daily Telegraph

Manufacturers Record

McClure's Magazine

Miami News

Monographs of Efficiency

Motor Age

Municipal Journal

New York Times

News Herald (Lawrenceville, Ga.)

North Georgia Citizen (Dalton)

Popular Science Monthly

Roads and Streets

Rome News-Tribune

Rome Tribune-Herald

Southern Good Roads

Southern Planter

Tennessee Agriculture

Thomaston Times

Waycross Journal-Herald

Worcester Magazine

Published Works

Abstract of Fourteenth Census of the United States. Washington, D.C.: Government Printing Office, 1923.

Acts and Resolutions of the General Assembly of the State of Georgia. Atlanta: Franklin Turner Company, 1907.

American Association for Highway Improvement. *Proceedings of the American Road Congress, Part II, Third to Sixth Days, Inclusive, September 30–October 5, 1912, Atlantic City, N.J.* Baltimore: Waverly Press, 1913.

Bulletin No. 24 of the Georgia Geological Survey. Atlanta: Charles P. Byrd, State Printer, 1910.

Chalmers, Hugh. "Relation of the Automobile Industry to the Good Roads Movement." *Proceedings of the American Road Congress, Richmond, Va., November 20–23, 1911,* 142–49. Baltimore: Waverly Press, 1912.

Compendium of the Tenth Census of the United States, part 1. Washington, D.C.: Government Printing Office, 1883.

Congressional Record.

First Annual Report of the State Highway Commission of Alabama. Montgomery: Brown Printing Company, 1912.

Fourteenth Census of the United States Taken in the Year 1920. Vol. 1, *Population.* Washington, D.C.: Government Printing Office, 1921.

Garrard, Louis F., and Henry R. Goetchius. *Revised Compilation of the Road Laws of the State of Georgia.* 5th ed. Columbus: Thos. Gilbert, Printer, 1886.

Geological Survey of Georgia. "A Second Report on the Public Roads of Georgia." Bulletin no. 24. Atlanta: Charles P. Byrd, 1910.

Georgia Department of Agriculture. *Georgia: The Empire State of the South: What She Is and Will Be.* Serial no. 65 (April 1915).

Georgia's General Assembly of 1880–1: Biographical Sketches of Senators, Representatives, the Governor and Heads of Departments. Atlanta: Jas. P. Harrison and Company, 1882.

Joy, Henry B. "To the Members of the National Association of Manufacturers." Detroit: n.p., 1909.

"Lincoln Highway Today, Summarizing a Year's Success, The." Detroit: Lincoln Highway Association, 1914.

Marker, James R. "Experimental Road Work in Ohio." *Annals of the American Academy of Political and Social Science* 46, *Prison Labor* (March 1913): 97–98.

McCallie, J. W. "A Second Report on the Public Roads of Georgia." Bulletin no. 24 of the Georgia Geological Survey. Atlanta: Charles P. Byrd, State Printer, 1910.

Official Automobile Blue Book. New York: Automobile Blue Book Publishing Co., 1901.

Potter, Isaac B. *The Gospel of Good Roads: A Letter to the American Farmer.* New York: Evening Post Job Printing House, 1891.

Pratt, Joseph Hyde. "Convict Labor in Highway Construction." *Annals of the American Academy of Political and Social Science* 46, "Prison Labor" (March 1913): 78–87.

Proceedings of the Fourth American Road Congress, Atlanta, Georgia, November 9–14, 1914. Baltimore: Waverly Press, 1915.

Sheffield, O. H. "Improvement of the Road System of Georgia." U.S. Department of Agriculture, Office of Road Inquiry Bulletin No. 3. Washington, D.C.: Government Printing Office, 1894.

Statistics of the Population of the United States at the Tenth Census, 1880. Washington, D.C.: Government Printing Office, 1883. Reprint, New York: Norman Ross Publishing, 1991.

Thirteenth Census of the United States (1910): Abstract of the Census. Washington, D.C.: Government Printing Office, 1913.

Thirteenth Census of the United States Taken in the Year 1910. Vol. 2, *Population.* Washington, D.C.: Government Printing Office, 1913.

Twelfth Census of the United States, Taken in the Year 1900. Part 1, *Population.* Washington, D.C.: U.S. Census Office, 1901.

Tynan, Thomas J. "Prison Labor on Public Roads." *Annals of the American Academy of Political and Social Science* 46 (March 1913): 58–60.

U.S. Department of Agriculture. *Proceedings of the North Carolina Good Roads Convention, February 1902.* Bulletin no. 34. Washington, D.C.: Government Printing Office, 1903.

U.S. Department of Agriculture, Office of Road Inquiry. "Improvement of the Road System of Georgia." Bulletin No. 3. Washington, D.C.: Government Printing Office, 1894.

U.S. Department of Agriculture, Office of Public Roads. "Mileage and Cost of Public Roads in the United States in 1909." Bulletin no. 41. Washington, D.C.: Government Printing Office, 1912.

U.S. Department of Agriculture, Office of Public Road Inquiry. "The Railroad and the Wagon Road." Circular no. 37. Washington, D.C.: Government Printing Office, 1904.

U.S. Senate Committee on Agriculture and Forestry. *Roads and Road Building Hearing, 26 January, 1904.* Washington, D.C.: Government Printing Office, 1904.

United States Statues at Large.

Secondary Sources

Adams, Arthur Barto. "Marketing Perishable Farm Products." Ph.D. diss., Columbia University, 1916.

Aiken, Charles S. *The Cotton Plantation South since the Civil War.* Baltimore: The Johns Hopkins University Press, 2003.

American Association of State Highway Officials. *AASHO: The First Fifty Years, 1914–1964.* Washington, D.C., 1965.

Anderson, William. *The Wild Man from Sugar Creek: The Political Career of Eugene Talmadge.* Baton Rouge: Louisiana State University Press, 1975.

Aycock, Roger. *All Roads Lead to Rome.* Roswell, Ga.: W. H. Wolfe Associates, 1981.

Ayers, Edward L. *The Promise of the New South: Life after Reconstruction*. New York: Oxford University Press, 1992.

Baker, Robert S. *Chattooga: The Story of a County and Its People*. Roswell, Ga.: W. H. Wolfe Associates, 1988.

Balogh, Brian. *A Government Out of Sight: The Mystery of National Authority in Nineteenth-Century America*. New York: Cambridge University Press, 2009.

Bartley, Numan V. *The Creation of Modern Georgia*. Athens: University of Georgia Press, 1983.

Battey, George Magruder, Jr. *A History of Rome and Floyd County*. Vol. 1. Atlanta: Webb and Vary Company, 1922.

Belasco, Warren James. *Americans on the Road: From Autocamp to Motel, 1910–1945*. Baltimore: The Johns Hopkins University Press, 1979.

Berger, Michael L. *The Devil Wagon in God's Country: The Automobile and Social Change in Rural America, 1893–1929*. Hamden, Conn.: Archon Books, 1979.

Bonner, James C. *Milledgeville: Georgia's Antebellum Capital*. Athens: University of Georgia Press, 1978.

Brinkley, Alan. *Voices of Protest: Huey Long, Father Coughlin, and the Great Depression*. New York: Knopf, 1982.

Burk, Robert F. *The Corporate State and the Broker State: The Du Ponts and American National Politics, 1925–1940*. Cambridge, Mass.: Harvard University Press, 1990.

Burns, Robert. *I Am a Fugitive from a Georgia Chain Gang!* Edited by Matthew J. Mancini. Athens: University of Georgia Press, Brown Thrasher Books, 1997.

Carlton, David. *Mill and Town in South Carolina, 1880–1920*. Baton Rouge: Louisiana State University Press, 1982.

Carter, Charles Frederick. *When Railroads Were New*. New York: Henry Holt and Company, 1909.

Caudill, Edward, and Paul Ashdown. *Sherman's March in Myth and Memory*. Lanham, Md.: Rowman & Littlefield, 2009.

Chandler, Alfred D., Jr., and Stephen Salsbury. *Pierre S. Du Pont and the Making of the Modern Corporation*. New York: Harper & Row, 1971.

Chernow, Ron. *The House of Morgan: An American Banking Dynasty and the Rise of Modern Finance*. New York: Grove Press, 1990.

Childs, William R. *Trucking and the Public Interest: The Emergence of Federal Regulation, 1914–1940*. Knoxville: University of Tennessee Press, 1985.

Coleman, Kenneth, ed. *A History of Georgia*. Athens: University of Georgia Press, 1991.

Cox, Karen L. *Dreaming of Dixie: How the South Was Created in American Popular Culture*. Chapel Hill: University of North Carolina Press, 2011.

Daniel, Pete. *Breaking the Land: The Transformation of Cotton, Tobacco, and Rice Cultures since 1880*. Urbana: University of Illinois Press, 1985.

———. *Lost Revolutions: The South in the 1950s*. Chapel Hill: University of North Carolina Press, 2000.

Davies, Pete. *American Road: The Story of an Epic Transcontinental Journey at the Dawn of the Motor Age*. New York: Henry Holt and Company, 2002.

Dearing, Charles L. *American Highway Policy*. Washington, D.C.: Brookings Institution, 1941.

Desmond, Jerry R. *Chattanooga*. Charleston, W.Va.: Arcadia Publishing, 1996.

Dinnerstein, Leonard. *The Leo Frank Case*. New York: Columbia University Press, 1968.

Dittmer, John. *Black Georgia in the Progressive Era, 1900–1920*. Urbana: University of Illinois Press, 1977.

Doyle, Don H. *New Men, New Cities, New South: Atlanta, Nashville, Charleston, Mobile, 1860–1910*. Chapel Hill: University of North Carolina Press, 1990.

Draper, Arthur Stimson, and others. "Michigan's Great Booze Rush and Its Suppression by State and Federal Action." *Literary Digest*, March 15, 1919, 85–90.

Dunkelman, Mark H. *Marching with Sherman: Through Georgia and the Carolinas with the 154th New York*. Baton Rouge: Louisiana State University Press, 2012.

Durden, Marion Little. *A History of Saint George Parish, Colony of Georgia, Jefferson County, State of Georgia*. Swainsboro, Ga.: Magnolia Press, 1983.

Eisenhower, Dwight D. *At Ease: Stories I Tell to Friends*. New York: Doubleday and Company, Inc., 1967.

Engelmann, Larry. *Intemperance: The Lost War against Liquor*. New York: Free Press, 1979.

Faris, John T. *Roaming American Highways*. New York: Farrar and Rinehart, 1931.

Fisher, Jane. *Fabulous Hoosier: A Story of American Achievement*. New York: Robert M. McBride & Co., 1947.

Fisher, Jerry M. *The Pacesetter: The Untold Story of Carl G. Fisher*. Fort Bragg, Calif.: Lost Coast Press, 1998.

Fite, Gilbert C. *Cotton Fields No More: Southern Agriculture, 1865–1980*. Lexington: University of Kentucky Press, 1984.

———. *Richard B. Russell, Jr., Senator from Georgia*. Chapel Hill: University of North Carolina Press, 1991.

Flamming, Douglas. *Creating the Modern South: Millhands and Managers in Dalton, Georgia, 1884–1984*. Chapel Hill: University of North Carolina Press, 1992.

Flanagan, Maureen A. *America Reformed: Progressives and Progressivisms, 1890–1920*. New York: Oxford University Press, 2007.

Flink, James T. *America Adopts the Automobile, 1895–1910*. Cambridge, Mass.: MIT Press, 1970.

———. *The Automobile Age*. Cambridge, Mass.: MIT Press, 1988.

———. *The Car Culture*. Cambridge, Mass.: MIT Press, 1975.

Ford, Henry. *My Life and Work*. New York: Garden City Publishing, 1922.

Formwalt, Lee. "A Garden of Irony and Diversity." In *The New Georgia Guide*, 497–536. Athens: University of Georgia Press, 1996.

Foster, Mark S. *Castles in the Sand: The Life and Times of Carl Graham Fisher*. Gainesville: University Press of Florida, 2000.

Francis, William, and Michael C. Hahn. *The DuPont Highway*. Charleston, S.C.: Arcadia Publishing, 2009.

Friend, Craig Thompson. *Along the Maysville Road: The Early American Republic in the Trans-Appalachian West*. Knoxville: University of Tennessee Press, 2005.

Fuller, Wayne E. "The South and the Rural Free Delivery of Mail." *Journal of Southern History* 25, no. 4 (November 1959): 499–521.

Fussell, Fred C. "Touring Southwest Georgia." In *The New Georgia Guide*, 537–61. Athens: University of Georgia Press, 1996.

Galloway, Tammy Harden. "'Tribune of the Masses and Champion of the People': Eugene Talmadge and the Three-Dollar Tag." *Georgia Historical Quarterly* 79, no. 3 (Fall 1995): 673–84.

Georgia Humanities Council. *The New Georgia Guide*. Athens: University of Georgia Press, 1996.

Giesen, James C. *Boll Weevil Blues: Cotton, Myth, and Power in the American South*. Chicago: University of Chicago Press, 2011.

Gilmore, Glenda. *Gender and Jim Crow: Women and the Politics of White Supremacy in North Carolina, 1896–1920*. Chapel Hill: University of North Carolina Press, 1996.

Goddard, Stephen B. *Getting There: The Epic Struggle between Road and Rail in the American Century*. New York: Basic Books, 1994.

Goodwyn, Lawrence. *The Populist Moment: A Short History of the Agrarian Revolt in America*. New York: Oxford University Press, 1978.

Gordon, Sarah. *Passage to Union: How the Railroads Transformed American Life, 1829–1929*. Chicago: Ivan R. Dee, 1996.

Govan, Gilbert E., and James W. Livingood. *The Chattanooga Country, 1540–1976: From Tomahawks to TVA*. 3rd ed. Knoxville: University of Tennessee Press, 1977.

Grantham, Dewey W., Jr. "Hoke Smith: Progressive Governor of Georgia, 1907–1909." *Journal of Southern History* 15 (November 1949): 423–40.

———. *Hoke Smith and the Politics of the New South*. Baton Rouge: Louisiana State University Press, 1958.

Guerry, Alexander, Jr. *Men and Vision: The Secret of Yesterday's Success, the Formula for Tomorrow's: A Brief History of the Chattanooga Medicine Company*. New York: Newcomen Society in North America, 1963.

Gutfreund, Owen D. *Twentieth-Century Sprawl: Highways and the Reshaping of the American Landscape*. New York: Oxford University Press, 2004.

Hackbart-Dean, Pamela. "Georgia's Renaissance Governor: Lamartine Hardman—Physician, Millowner, Agriculturalist." *Georgia Historical Quarterly* 79 (Summer 1995): 441–52.

Hahamovitch, Cindy. *The Fruits of Their Labor: Atlantic Coast Farmworkers and the Making of Migrant Poverty, 1870–1945*. Chapel Hill: University of North Carolina Press, 1997.

Hair, William Ivy. *The Kingfish and His Realm: The Life and Times of Huey P. Long*. Baton Rouge: Louisiana State University Press, 1991.

Hall, Jacqueline Dowd, James Leloudis, Robert Korstad, Mary Murphy, Christopher B. Daly, and Lu Ann Jones. *Like a Family: The Making of a Southern Cotton Mill World*. Chapel Hill: University of North Carolina Press, 1987.

Hall, Randall L. "Before NASCAR: The Corporate and Civic Promotion of Automobile Racing in the American South, 1903-1927." *Journal of Southern History* 68 (August 2002): 629-68.

Hamilton, Shane. *Trucking Country: The Road to America's Wal-Mart Economy*. Princeton, N.J.: Princeton University Press, 2008.

Herlihy, David V. *Bicycle: The History*. New Haven, Conn.: Yale University Press, 2006.

Hilles, William C. "The Good Roads Movement in the United States, 1880-1916." M.A. thesis, Duke University, 1958.

Hokanson, Drake. *The Lincoln Highway: Main Street across America*. Iowa City: University of Iowa Press, 1988.

Holley, I. B., Jr. *The Highway Revolution, 1895-1925: How the United States Got out of the Mud*. Durham, N.C.: Carolina Academic Press, 2008.

Holmes, Yulssus Lynn. *Those Glorious Days: A History of Louisville as Georgia's Capital, 1796-1807*. Macon, Ga.: Mercer University Press, 1996.

Host, William R., and Brooke Ahne Portmann. *Early Chicago Hotels*. Chicago: Arcadia Publishing, 2006.

Hudson, Angela Pulley. *Creek Paths and Federal Roads: Indians, Settlers, and Slaves and the Making of the American South*. Chapel Hill: University of North Carolina Press, 2010.

Hugill, P. J. "Good Roads and the Automobile in the United States, 1880-1929." *Geographical Review* 72, no. 3 (1982): 327-49.

Hulbert, Archer B. *The Cumberland Road*. Cleveland: Arthur H. Clark Company, 1904.

———. *The Paths of Inland Commerce: A Chronicle of Trail, Road and Waterway*. New Haven: Yale University Press, 1921.

Hulbert, Archer B. *The Future of Road-Making in America*. Cleveland: Arthur H. Clark Company, 1905.

Hurst, Robert Latimer. *This Magic Wilderness: Part I and Part II*. Waycross, Ga.: Wilderness Publications, 1982.

Ingram, Tammy. "Distribution before Diversification: Good Farms, Bad Roads, and Agricultural Reform in South Carolina in the 1920s." Unpublished conference paper delivered to the Southern Historical Association, November 2005.

Jackson, Kenneth T. *The Ku Klux Klan in the City, 1915-1930*. New York: Oxford University Press, 1970.

Jakle, John A., and Keith A. Sculle. *Motoring: The Highway Experience in America*. Athens: University of Georgia Press, 2008.

Jones, Lu Ann. *Mama Learned Us to Work: Farm Women in the New South*. Chapel Hill: University of North Carolina Press, 2002.

Karnes, Thomas L. *Asphalt and Politics: A History of the American Highway System*. Jefferson, N.C.: McFarland & Company, Inc., 2009.

Keith, Jeanette. "Lift Tennessee out of the Mud: Ideology and the Good Roads Movement in Tennessee." *Southern Historian* 9 (1988): 22–37.

Kennedy, David M. *Over Here: The First World War and American Society.* New York: Oxford University Press, 1980.

Kirby, Jack Temple. *Rural Worlds Lost: The American South, 1920–1960.* Baton Rouge: Louisiana State University Press, 1987.

Knight, Lucian Lamar. *Georgia's Landmarks, Memorials, and Legends.* Vol. 1. Byrd Printing Company, 1913.

Kuhn, Clifford M. *Contesting the New South Order: The 1914–1915 Strike at Atlanta's Fulton Mills.* Chapel Hill: University of North Carolina Press, 2001.

Kytle, Calvin, and James A. Mackay. *Who Runs Georgia?* Athens: University of Georgia Press, 1998.

Leach, William. *Land of Desire: Merchants, Power, and the Rise of a New American Culture.* New York: Vintage Books, 1993.

Lesseig, Corey T. "'Out of the Mud': The Good Roads Crusade and Social Change in Twentieth-Century Mississippi." *Journal of Mississippi History* 60, no. 1 (1998): 50–72.

Leynes, Jennifer Brown, and David Cullison. *Biscayne National Park: Historic Resource Study.* Atlanta: National Park Service Southeast Regional Office, 1998.

———. "Recreational Development of Miami and Biscayne Bay, 1896-1945." In *Biscayne National Park: Historic Resource Study*, 19–37. Atlanta: National Park Service Southeast Regional Office, 1998.

Lichtenstein, Alex. "Good Roads and Chain Gangs in the Progressive South: 'The Negro Convict Is a Slave.'" *Journal of Southern History* 59 (February 1993): 85–110.

———. *Twice the Work of Free Labor: The Political Economy of Convict Labor in the New South.* New York: Verso, 1996.

Link, William A. *The Paradox of Southern Progressivism, 1880–1930.* Chapel Hill: University of North Carolina Press, 1992.

Lomax, John A., and Alan Lomax. *American Ballads and Folk Songs.* New York: Dover Publications, Inc., 1994.

Love, Steve, and David Giffels. *Wheels of Fortune: The Story of Rubber in Akron.* Akron: University of Akron Press, 1999.

Lowry, Amy Gillis, and Abbie Tucker Parks. *North Georgia's Dixie Highway.* Charleston, W.Va.: Arcadia Publishing, 2007.

Madsen, Axel. *Deal Maker: How William C. Durant Made General Motors.* New York: John Wiley and Sons, 1999.

Mancini, Matthew J. *One Dies, Get Another: Convict Leasing in the American South, 1866–1928.* Columbia: University of South Carolina Press, 1996.

McDaniel, Susie Blaylock. *Official History of Catoosa County, Georgia, 1853–1953.* Dalton, Ga.: Gregory Printing and Office Supply, 1953.

McGerr, Michael. *A Fierce Discontent: The Rise and Fall of the Progressive Movement in America, 1870–1920.* New York: Oxford University Press, 2003.

McKenzie, Roderick Clayton. "The Development of Automobile Road Guides in the United States." M.A. thesis, University of California at Los Angeles, 1963.

McKown, Harry Wilson, Jr. "Roads and Reform: The Good Roads Movement in North Carolina, 1885–1921." M.A. thesis, University of North Carolina at Chapel Hill, 1972.

Melton, Brian. "The Town That Sherman Wouldn't Burn: Sherman's March and Madison, Georgia, in History, Memory, and Legend." *Georgia Historical Quarterly* 86, no. 2 (Summer 2002): 201–30.

National Highway Users Conference. *Texts of Good Roads Amendments: State Constitutional Provisions Safeguarding Highway User Taxes.* Washington, D.C.: NHUC, 1949.

Nelson, Megan Kate. *Trembling Earth: A Cultural History of the Okefenokee Swamp.* Athens: University of Georgia Press, 2005.

Nelson, Scott Reynolds. *Steel Drivin' Man: John Henry, the Untold Story of an American Legend.* New York: Oxford University Press, 2006.

Newton, Michael. *The Ku Klux Klan: History, Organization, Language, Influence, and Activities of America's Most Notorious Secret Society.* Jefferson, N.C.: McFarland & Company, Inc., 2007.

Okrent, Daniel. *Last Call: The Rise and Fall of Prohibition.* New York: Scribner, 2007.

Oney, Steve. *And the Dead Shall Rise: The Murder of Mary Phagan and the Lynching of Leo Frank.* New York: Pantheon Books, 2003.

O'Reilly, Maurice. *The Goodyear Story.* Elmsford, N.Y.: Benjamin Company, Inc., 1983.

Oshinsky, David M. *"Worse than Slavery": Parchman Farm and the Ordeal of Jim Crow Justice.* New York: Free Press, 1996.

Ownby, Ted. *American Dreams in Mississippi: Consumers, Poverty, and Culture, 1830–1998.* Chapel Hill: University of North Carolina Press, 1999.

Perman, Michael. *Struggle for Mastery: Disfranchisement in the South, 1888–1908.* Chapel Hill: University of North Carolina Press, 2001.

Phillips, Clifton J. *Indiana in Transition: The Emergence of an Industrial Commonwealth, 1880–1920.* Indianapolis: Indiana Historical Bureau and Indiana Historical Society, 1968.

Pierce, Daniel S. *Real NASCAR: White Lightning, Red Clay, and Big Bill France.* Chapel Hill: University of North Carolina Press, 2010.

Postel, Charles. *The Populist Vision.* New York: Oxford University Press, 2009.

Preston, Howard Lawrence. *Dirt Roads to Dixie: Accessibility and Modernization in the South, 1885–1935.* Knoxville: University of Tennessee Press, 1991.

Rae, John B. *The American Automobile: A Brief History.* Chicago: University of Chicago Press, 1965.

Range, Willard. *A Century of Georgia Agriculture, 1850–1950.* Athens: University of Georgia Press, 1954.

Ray, Janisse. *Ecology of a Cracker Childhood.* Minneapolis: Milkweed Editions, 1999.

Rigdon, Louis T. *Georgia's County Unit System.* Decatur, Ga.: Selective Books, 1961.

Rogers, William Warren. *Thomas County, 1865–1900.* Tallahassee: Florida State University Press, 1973.

Rose, Mark H. *Interstate: Express Highway Politics, 1939–1989.* Knoxville: University of Tennessee Press, 1990.

Rose, Mark H., Bruce E. Seely, and Paul F. Barrett. *The Best Transportation System in The World: Railroads, Trucks, Airlines, and American Public Policy in the Twentieth Century*. Philadelphia: University of Pennsylvania Press, 2006.

Ross, Stewart Halsey. *Propaganda for War: How the United States Was Conditioned to Fight the Great War of 1914–1918*. McFarland & Co., Inc., 1996.

Sanders, Elizabeth. *Roots of Reform: Farmers, Workers, and the American State, 1877–1917*. Chicago: University of Chicago Press, 1999.

Savitt, Todd L. *Disease and Distinctiveness in the American South*. Knoxville: University of Tennessee Press, 1988.

Seely, Bruce. *Building the American Highway System: Engineers as Policy Makers*. Philadelphia: Temple University Press, 1987.

Shaffer, Marguerite S. *See America First: Tourism and National Identity, 1880–1940*. Washington, D.C.: Smithsonian Institution Press, 2001.

Sharp, Leslie. *Tennessee's Dixie Highway: Springfield to Chattanooga*. Charleston, W.Va.: Arcadia Publishing, 2011.

Simon, Bryant. *A Fabric of Defeat: The Politics of South Carolina Millhands, 1910–1948*. Chapel Hill: University of North Carolina Press, 1998.

Skocpol, Theda. "Political Responses to Capitalist Crisis: Neo-Marxist Theories of the State and the Case of the New Deal." *Politics and Society* 10 (1980): 155–202.

———. *Social Policy in the United States: Future Possibilities in Historical Perspective*. Princeton, N.J.: Princeton University Press, 1995.

Skowronek, Stephen. *Building a New American State: The Expansion of National Administrative Capacities, 1877–1920*. Cambridge: Cambridge University Press, 1982.

Smith, Lillian. "Memory of a Large Christmas." *Life*, December 15, 1961.

Southerland, Henry de Leon, Jr., and Jerry Elijah Brown. *The Federal Road through Georgia, the Creek Nation, and Alabama, 1806–1836*. Tuscaloosa: University of Alabama Press, 1989.

Stager, Claudette, and Martha Carver, eds. *Looking beyond the Dixie Highway: Dixie Roads and Culture*. Knoxville: University of Tennessee Press, 2006.

Steiner, Jesse F., and Roy M. Brown. *The North Carolina Chain Gang: A Study of County Convict Road Work*. Chapel Hill: University of North Carolina Press, 1927.

Stover, John F. *American Railroads*. Chicago: University of Chicago Press, 1961.

———. *The Railroads of the South, 1865–1900: A Study in Finance and Control*. Chapel Hill: University of North Carolina Press, 1955.

Sullivan, Buddy. *Georgia: A State History*. Charleston, W.Va.: Arcadia Publishing, 2010.

Sullivan, Larry E. *The Prison Reform Movement: Forlorn Hope*. Boston: Twayne Publishers, 1990.

Summers, Mark W. *Railroads, Reconstruction, and the Gospel of Prosperity: Aid under the Radical Republicans, 1865–1877*. Princeton, N.J.: Princeton University Press, 1984.

Sutter, Paul S. "Paved with Good Intentions: Good Roads, the Automobile, and the Rhetoric of Rural Improvement in the *Kansas Farmer*, 1890–1914." *Kansas History* 18, no. 4 (Winter 1995–96): 284–99.

Teaford, Jon C. *The Rise of the States: Evolution of American State Government.* Baltimore: The Johns Hopkins University Press, 2002.

Tindall, George B. *The Emergence of the New South, 1913–1945.* Baton Rouge: Louisiana State University Press, 1951.

U.S. Department of the Interior. *Laws Relating to the National Park Service: Supplement II, May 1944 to January 1963.* Washington, D.C.: Government Printing Office, 1963.

U.S. Department of Transportation, Federal Highway Administration. *America's Highways, 1776–1976: A History of the Federal Aid Program.* Washington, D.C.: Government Printing Office, 1976.

Vickers, Raymond B. *Panic in Paradise: Florida's Banking Crash of 1926.* Tuscaloosa: University of Alabama Press, 1994.

Watts, Steven. *The People's Tycoon: Henry Ford and the American Century.* New York: Vintage, 2005.

Weingroff, Richard F. "Federal Aid Road Act of 1916: Building the Foundation." *Public Roads* 60, no. 1 (Summer 1996): 2–6.

———. "From Names to Numbers: The Origins of the Numbered U.S. Highway System." *AASHTO Quarterly* (Spring 1997).

Wells, Christopher W. "The Changing Nature of Country Roads: Farmers, Reformers, and the Shifting Uses of Rural Space, 1880–1905." *Agricultural History* 80 (Spring 2006): 143–66.

———. "Fueling the Boom: Gasoline Taxes, Invisibility, and the Growth of the American Highway Infrastructure." *Journal of American History* 99 (June 2012): 72–81.

Wetherington, Mark. *The New South Comes to Wiregrass Georgia, 1860–1910.* Knoxville: University of Tennessee Press, 1994.

Whitfield-Murray Historical Society. *An Official History of Whitfield County, Georgia, 1852–1999.* Fernandina Beach, Fla.: Wolfe Publishing, 1999.

Williford, William Bailey. *Americus through the Years: The Story of a Georgia Town and Its People, 1832–1975.* Atlanta: Cherokee Publishing Company, 1975.

Woodward, C. Vann. *Origins of the New South, 1877–1913.* Baton Rouge: Louisiana State University Press, 1951.

———. *Tom Watson: Agrarian Rebel.* New York: Oxford University Press, 1938.

Wright, Gavin. *Old South, New South: Revolutions in the Southern Economy since the Civil War.* New York: Basic Books, 1986.

Wright, James R. *The Dixie Highway in Illinois.* Charleston, W.Va.: Arcadia Publishing, 2009.

Wrone, David R. "Illinois Pulls out of the Mud." *Journal of the Illinois State Historical Society* 58, no. 1 (Spring 1965): 54–76.

Index

Cumberland Mountain, Tenn., 120
Cumberland Road, 19

Daimler, Gottleib, 24
Dalton, Ga., 6, 59–64, 67, 81–82, 109, 171, 189
Danville, Ill., 55
Davis, Scott, 142
Deese, Joel, 36–37
Democratic-Republicans, 18–19
Dempsey, Jack, 77
Diehl, George, 41, 99, 122
Disfranchisement. *See* African American Disfranchisement
Dixie Highway: origins of, 1–2, 31–32, 43–52; and sectional unity, 43–44; and governors conference, 48, 51, 54, 75–79, 81, 83, 136; federal aid for, 48–49, 51–52, 85, 87–89, 105, 119–22, 172–73 (*see also* World War I: Dixie Highway Association propaganda during); state aid for, 48–49, 51–52, 119–22, 172–73; local business support for, 49–50, 102–8 (*see also* World War I: trucking); routing contest for, 53–77, 80–82, 111, 169, 178; and tourism, 57, 61–65, 67–68, 70–72, 101, 105–12, 125 (*see also* "See America First" campaign); selection of official route of, 80–84; local government aid for, 85, 118–20; official maps of, 112–14; signage of, 114, 165, 174–75; legacy of, 188–92, 196, 198–99
Dixie Highway (magazine), 96, 98–100, 104–5, 107–10, 115–16, 119, 121–23, 125–27, 172
Dixie Highway Association, 2, 7–9, 11–12, 89, 91, 112, 131–32, 165, 168, 178–79, 186; support for federal aid, 44–45, 84–86, 116, 118–22, 127, 144–45, 147, 161 (*see also* Dixie Highway: federal aid for; World War I: Dixie Highway Association propaganda during); origins of, 76–80, 83; support for state highway

bonds, 148–49, 155, 159; opposition to local control, 149; end of, 172, 175, 188
Dixie Portland Cement Company, 77, 107–8
Dodge, Martin, 22, 34
Donalsonville, Ga., 170
Dorsey, Hugh M., 18
Dougherty County, Ga., 36. *See also* Albany, Ga.
DuPont, Coleman, 77
Durant, William C. "Billy," 47

Eisenhower, Dwight D., 3, 10–11, 91–92, 125, 198
Eisenhower interstate system, 4, 11, 196, 198–99
Ellijay, Ga., 36

Farmers' Alliance, 20. *See also* Railroads: farmers' hostility toward
Federal highway legislation, 34, 41, 49–51, 85–87, 189–92, 196–98; Post Office Appropriation Bill of 1913, 3; Brownlow-Latimer Bill, 34; Shackleford Bill, 34; Bankhead-Shackleford Bill, 87–88; divisions over, 87–88, 151–52; Federal Aid Road Act of 1916, 88–89, 92–93, 95–96, 98–100, 118–19, 122, 124–25, 143–45, 147–53, 163, 165, 170; Chamberlain-Dent highway bill, 116–17; Townsend Bill, 124, 151; Federal Aid Highway Act of 1921, 131, 152–55, 157–58, 163, 165, 168; Federal Aid Highway Act of 1934, 196–97; Federal Aid Highway Act of 1956 (*see* Eisenhower, Dwight D.). *See also* Highway lobby
Federal highway system, 86–87, 124, 148, 151, 168–69, 172; 1926 U.S. Highway System, 164, 169–71, 175–79, 188–92; and Dixie Highway, 171, 177
Federalists, 18
Federal Road, 19
Finnegan, Richard J., 78–79
Firestone Tire and Rubber Company, 47

Fisher, Carl Graham, 7, 28–31, 46, 85, 116–17, 149, 163; role of in creating Dixie Highway, 43, 45, 47–51, 75–78, 84, 196

Fisher, Jane, 29, 47

Fitzgerald, Ga., 35, 49, 105, 144

Flint River, 69

Florida, 10, 47, 77, 125; state highway department of, 51, 120; Dixie Highway in, 54, 58, 70, 81–82, 85, 106, 120, 144

Florida Short Route, 170

Floyd County, Ga., 64, 119, 141–42

Floyd County Board of Roads and Revenues, 141–42

Ford, Edsel, 47

Ford, Henry, 24, 26, 28; and Model T, 24, 26–28

Fort McPherson, 91

Fort Oglethorpe, 91

Fort Screven, 100

Fort Thomas, 101

Fort Valley, Ga., 69

French Lick Springs, Ind., 78

Fruit Growers Exchange, 69–70

Fulton County, Ga., 35, 61. *See also* Atlanta, Ga.

Fyffe, Colonel J. P., 96–97

Gasoline taxes, 131, 156–57, 171, 179, 181–82, 184, 186, 192, 194–95, 197, 200

General Motors, 47

Georgetown, Ga., 67

Georgia: Dixie Highway in, 10, 48, 52, 54, 58–74, 81–82, 85, 111, 118, 120, 143–45, 153–55, 157–60, 172, 188–89, 192, 199; road and highway mileage in, 10, 52, 58, 143–44, 153–55, 157, 160, 181, 198; 1926 gubernatorial race, 12, 164, 171, 177, 179–88, 192–93 (*see also* Holder, John; Hardman, Lamartine G.); good-roads clubs in, 23; automobile sales in, 32; state highway department of, 120, 143, 147, 149–51, 154–61, 178–86, 193–95 (*see also* Neel,

Warren R.); state highway system in, 146, 153–56, 172, 181–82, 184, 186; state highway revenues in, 149–50, 153–60, 171, 180–86, 192–94, 198, 200. *See also* County unit system

Georgia and Alabama Railroad, 132–33

Georgia-Florida Pecan Growers Association, 69

Georgia Institute of Technology, 145

Gilbreath, William S., 48–49, 53–54, 57, 75, 84, 86

Glidden, Charles J., 38

Glidden Tours, 38–39, 41

Golden Gate Bridge, 9

Good Roads, 21

Good Roads Movement, 1, 3–4, 7, 10–11, 28, 33, 35–36, 41–42, 44, 48, 52, 84–85, 87–89, 91–92, 95, 101–2, 113–15, 125, 127, 131, 134, 142–43, 147; origins of, 8, 20–21, 178, 193 (*see also* League of American Wheelmen); relationship between farmers and motorists in, 15, 23–25, 32, 40–42; clubs, 21, 23, 34, 116; divisions within, 151, 164, 173. *See also* Chain gangs: support for and opposition to

Good Roads Trains, 22

Goodyear Tire and Rubber Company, 28. *See also* Frank A. Seiberling

Gordon County, Ga., 63–64

Gore, Howard M., 168–69, 173

Gould, Jay, 20

Governors conference. *See* Dixie Highway: governors conference

Grady, Henry, 40

Great Depression, 3, 192, 194, 196–97

Great Lakes, 81

Great Migration, 136–37

Greenville, S.C., 165

Gulf of Mexico, 81

Hanger, Harry B., 78

Hardman, Lamartine G., 164, 177, 179–88, 192

Packard Motors, 47, 116
Page, Logan Waller, 34, 36, 38, 117, 165
Panama Canal, 9
Pan American Railroad, 145
Paterson, N.J., 104
Patten, John A., 77, 108
Patterson, T. E., 150–51
Paulding County, Ga., 172
Pay-as-you-go plan, 149–50, 156, 164, 171, 179, 183, 186
Peach industry, 69–70
"Peacock alley," 25, 41
Pecan industry, 69–71
Penn, J. C., 141–42
Philadelphia, Pa., 104
Pinehurst, N.C., 40
Pomerane, Atlee, 88
Pope, Colonel Albert, 24, 28
Populist Party, 20, 22. *See also* Railroads: farmers' hostility toward
Post, Emily, 46
Post roads. *See* Roads and Highways: post roads
Pratt, Joseph Hyde, 49, 136
Prest-O-Lite, 30
Preston, Howard Lawrence, 4
Progressive Era, 1–8; highway reform during, 2, 9, 11, 23; reform and regulation during, 25–26, 40–41, 52–53, 94; conservatism of reform in South during, 130–31, 133–36, 163–64. *See also* Chain gangs
Progressive Party, 31. *See also* Fisher, Carl
Progressivism. *See* Progressive Era
Prohibition, 6, 15, 135
Pure Food and Drug Act, 25, 77

Railroads, 5–6, 14; long-distance highways' challenges to, 15, 64–65, 91–92, 96–97; mileage of, 19, 133; farmers' hostility toward, 19–20; federal support for, 19–20; monopolies in, 19–20; regulation of, 20, 94 (*see also* Interstate Commerce Commission);
problems during World War I of, 94–98, 102
Raleigh, N.C., 175
Ralston, Samuel, 43, 47–49, 76, 78
Rand McNally, 113
Reconstruction, 52
Return Loads Bureau, 118
Rhode Island, 16
Richardson, Edmund, 132
Richmond, Ky., 78
Richmond, Va., 175
Rickenbacker, Eddie, 30
Ringgold Gap, Ga., 61
Road and highway maps, 112–13
Roads and highways: local control over, 1, 9, 11–12, 16, 31, 35–37, 51, 58–59, 63–64, 73–76, 89, 147, 153, 160–61, 164, 174, 183–84 (*see also* Chain gangs; Georgia: 1926 gubernatorial race); local or farm-to-market roads, 1–6, 14–15, 25, 31, 41, 74, 151; long-distance highways, 1–6, 15, 25, 31–33, 40–41, 43, 91, 151, 153 (*see also* Marked trails); federal support for, 2–3, 8, 32–34, 36–37; taxes for, 3, 16–17, 33, 73, 145; regional differences in, 9, 12, 16–18, 54, 59, 159–60, 165–66, 192; cost of, 12, 16; poor condition of early roads, 13–14, 39, 172; local bonds for, 16, 107, 145, 185 (*see also* State highway bonds); congressional authority to build, 18–19; debates over during the early national era, 18–19; toll roads, 19; post roads, 19, 22–23, 34, 61, 86–89, 92, 118, 122, 153; in Roman Empire, 93–94, 116; paved roads, 106–7, 154–55, 159–60
Roanoke, Va., 40
Robert E. Lee Highway, 44
Robinson, V. D. L., 97, 103
Robsion, John M., 151–52
Rockcastle County, Ky., 120
Rogers, Will, 77
Rome, Ga., 59–64, 81–82, 119, 171–72, 188–89. *See also* Floyd County, Ga.